TECHNOLOGY, SPORT AND RECREATION

This book critically examines the transformative intersection of technology with sport and recreation, revealing how technological innovations reshape performance, participation, community engagement, and sustainability.

Drawing from empirical research, diverse case studies, and industry insights, the book explores technologies such as wearable fitness devices, artificial intelligence (AI), augmented and virtual reality, esports platforms, digital media, and smart sport-tech solutions. Readers will gain insights into how technology enhances personal fitness and athletic performance, creates immersive spectator experiences, facilitates digital community-building, promotes inclusivity in adaptive sports, and drives sustainability in sport management. Structured around the innovative Technology–Sport–Recreation Framework (TSRF), the book systematically captures complex technological interactions within sport and recreation, highlighting both benefits and ethical considerations. Each chapter includes case studies, practitioner interviews, and links to online resources to further enrich readers' understanding and application of the concepts discussed.

This is an essential resource for sport and recreation professionals, policymakers, academics, and students in sport management, kinesiology, technology studies, public health, and sustainability. It is also invaluable for technology developers, industry practitioners, and anyone passionate about understanding the dynamic interplay between sport, recreation, and technological advancement.

Jacob Bustad is an associate professor in the Department of Kinesiology at Towson University, USA. His primary research and teaching interests are in the fields of sport management, physical cultural studies, the sociology of sport, and urban studies.

Gashaw Abeza is an associate professor in the Department of Kinesiology at Towson University, USA. His research interest is in marketing communication, particularly digital media in the context of sport.

TECHNOLOGY, SPORT AND RECREATION

Edited by Jacob Bustad and Gashaw Abeza

Routledge
Taylor & Francis Group
LONDON AND NEW YORK

Designed cover image: Getty Images / liuzishan

First published 2026
by Routledge
4 Park Square, Milton Park, Abingdon, Oxon OX14 4RN

and by Routledge
605 Third Avenue, New York, NY 10158

Routledge is an imprint of the Taylor & Francis Group, an informa business

British Library Cataloguing-in-Publication Data
A catalogue record for this book is available from the British Library

ISBN: 978-1-032-96182-8 (hbk)
ISBN: 978-1-032-96180-4 (pbk)
ISBN: 978-1-003-58840-5 (ebk)

DOI: 10.4324/9781003588405

Typeset in Times New Roman
by SPi Technologies India Pvt Ltd (Straive)

CONTENTS

CONTRIBUTORS

Artemisia Apostolopoulou, Robert Morris University, USA

Andrew Baerg, University of Houston-Victoria, USA

Armand Buzzelli, Robert Morris University, USA

John Clark, Robert Morris University, USA

Yoonki Chun, Bowling Green State University, USA

James Du, Florida State University, USA

Chad Goebert, Kennesaw State University, USA

Susmit S. Gulavani, Florida State University, USA

Robin Hardin, University of Tennessee, USA

Angela I. Hayslett, James Madison University, USA

Jeffrey D. James, Florida State University, USA

jay johnson, University of Manitoba, Canada

John David Johnson, Kennesaw State University, USA

Philip Kang, Augusta University, USA

Niveditha Karthikeyan, Florida State University, USA

Cindy Lee, West Virginia University, USA

Gary Lhotsky, West Virginia University, USA

Matthew A. Masucci, San Jose State University, USA

Katharine M. Nohr, Sports Futurists LLC, USA

Joshua R. Pate, Lees-McRae College, USA

Irina Perfilova, University of Tennessee, USA

Manning Zhang, Brandeis University, USA

1

TECHNOLOGY, SPORT, AND RECREATION

A Conceptual Framework

Gashaw Abeza and Jacob Bustad

Introduction

The intersection of technology, sport, and recreation (TSR) constitutes a dynamic and rapidly evolving field of scholarly inquiry. Central to understanding this intersection is an appreciation of the distinct yet interconnected roles these three domains play within contemporary society. Sport is conventionally understood as an organized, competitive physical activity governed by a set of rules and regulations designed to ensure fairness and measurable outcomes (Coakley, 2020). It encompasses structured competition ranging from amateur grassroots initiatives to highly commercialized professional sports (Abeza & Seguin, 2024). Importantly, sport serves broader societal functions, including fostering community cohesion, national pride, and cultural expression, significantly amplified by technological advancements in recent decades (Billings, Butterworth, & Turman, 2017). In a similar vein, recreation is typically characterized as leisure-based engagement, pursued voluntarily for personal enjoyment, relaxation, or intrinsic satisfaction (Hurd et al., 2023). Recreational activities span physical pursuits, such as hiking and cycling, as well as mentally stimulating or relaxing pastimes like gaming and digital leisure activities. Unlike organized sports, recreation prioritizes personal pleasure, relaxation, and social interaction rather than competitive outcomes (Hurd et al., 2023).

Technological innovations are increasingly redefining sport and recreational activities, among others, by enhancing accessibility (Lee & Uihlein, 2018), creating immersive digital experiences (Yadav & Pavlou, 2020), and facilitating social interactions through digital platforms (Glebova et al., 2025). As sport and recreation continue to evolve in tandem with technological advancements, such as wearable devices, immersive media, mobile applications, and online

DOI: 10.4324/9781003588405-1

communities, these developments have transformed how individuals participate in, interact with, and derive meaning from such activities. This transformation has been examined across various chapters of this book, each highlighting specific domains in which technology has reshaped practices, experiences, and infrastructures within sport and recreation. This convergence has given rise to a complex, multidimensional system in which technology functions not just as a supporting tool but as a central driver of innovation, accessibility, and user engagement. To more effectively understand the implications of this convergence, it is necessary to examine the components that characterize the interplay between technological systems and sport and recreation. This necessitates the visual representation of an ecosystem that delineates the key components of this triadic relationship. Accordingly, the following section introduces the TSR Ecosystem.

Technology, Sport, and Recreation Ecosystem

The TSR ecosystem represents (see Figure 1.1) a comprehensive structure highlighting the interconnected components critical for the integration of technology within the context of sport and recreation. This ecosystem comprises five core interdependent components: (1) technological infrastructure, (2) digital innovations, (3) sport and recreation activities, (4) stakeholders, and (5) management and governance, all of which collectively influence and shape *user experiences and outcomes*.

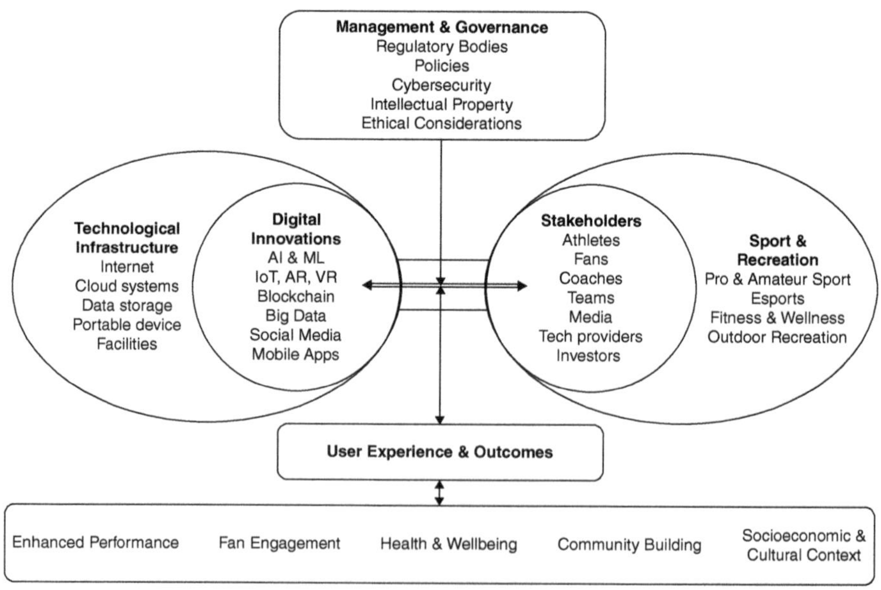

FIGURE 1.1 Technology, Sport, and Recreation Ecosystem.

The first component, *technological infrastructure*, provides the foundational elements necessary for technological deployment, including internet connectivity (e.g., Wi-Fi, 5G networks), cloud computing platforms (e.g., Amazon Web Services, Microsoft Azure), portable devices (e.g., smartphone, tablet, smartwatch), physical and digital facilities (e.g., virtual training platform, online gaming environment), and data storage and management systems (e.g., Amazon S3, Google Cloud Storage). As noted in different chapters of this book, technological infrastructure is vital for realizing the potential benefits of technological advancements in sport and recreation.

Digital innovations constitute the second component, characterized by cutting-edge technologies such as artificial intelligence (AI), machine learning (ML), the internet of things (IoT), augmented and virtual reality (AR and VR), blockchain, big data analytics, and social media platforms. As discussed in different chapters of the book, these innovations influence athletic performance, fan engagement, and personalized experiences in sport and recreational activities. The third component, *sport and recreation activities*, encompasses a broad spectrum of activities ranging from professional and amateur sports to esports, fitness, wellness, and outdoor recreational activities. As discussed across various chapters, technological integration within these activities has the potential to influence athletic performance, broader participation, and accessibility.

Stakeholders represent the fourth component, including athletes, spectators, coaches, technology developers, media outlets, and investors. These stakeholders collaboratively engage in potentially shaping the technological trajectory within sport and recreation settings, with their interactions being essential for the dynamic interrelationship between TSR. The fifth component, *management and governance*, addresses regulatory frameworks, policy development, intellectual property rights, cybersecurity, and ethical considerations. For example, as noted in Chapter 7, proper governance ensures ethical technology use, maintains compliance, and mitigates risks associated with data privacy and security.

Central to this ecosystem are *user experiences and outcomes*, which, as highlighted throughout various chapters of this volume, potentially include enhanced athletic performance, increased fan engagement, improved health and well-being, strengthened community connections, and broader socioeconomic and cultural benefits. Socioeconomic dimensions encompass aspects such as economic impacts, innovations in business models, and sustainable practices, while cultural dimensions involve diversity, globalization, and ethical considerations. Collectively, these components constitute an ecosystem with the potential to enrich experiences in sport and recreation, as well as to contribute to broader societal and economic development through the application of technology and interactions among stakeholders.

Anchored in the TSR ecosystem, this book examines the adoption, application, and implications of technologies (such as AI, VR, e-sports, digital fitness platforms, and smart sports equipment) within sport and recreation contexts. To locate the focus of this book within the broader ecosystem and systematically navigate the complex interplay among TSR, a structured approach is essential. Accordingly, the Technology–Sport–Recreation Framework (TSRF) is constructed as a conceptual model and is discussed in the following section.

Technology–Sport–Recreation Framework

The TSRF is a conceptual model that represents distinct yet interconnected aspects of technology adoption, practical application, and resulting impacts within the broader TSR ecosystem. The framework conceptualizes the intersection of TSR through four interconnected layers: *Performance Enhancement, Immersive Experiences, Digital Communities and Inclusion*, and *Smart Environments and Sustainability*. Each layer delineates specific yet interrelated dimensions of technological influence, emphasizing reciprocal rather than linear interactions. These layers directly correspond to the four main sections and individual chapters covered in this volume. Thus, the TSRF serves as a structured analytical tool to systematically examine and navigate the multifaceted relationships between technology and sport-recreation environments.

As depicted in Figure 1.2, the TSRF conceptualizes four interconnected layers, each representing distinct dimensions of technological mediation within sport and recreation contexts. These layers are structured along a continuum, illustrating increasing scales of technological mediation from the individual to societal levels. The first layer, *performance enhancement*, reflects direct physical engagement facilitated primarily through individualized technologies. This layer operates predominantly at the micro-level, emphasizing technologies aimed at optimizing individual athletic performance and participation. The second layer, *immersive experiences*, highlights digitally mediated participation and consumption via technologies such as VR and AR. Spanning micro- to meso-level interactions, this layer addresses immersive technological interfaces that transform experiences at individual and group levels.

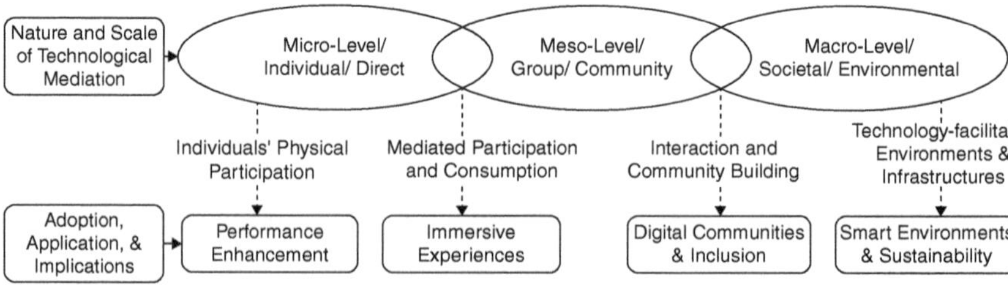

FIGURE 1.2 The Technology–Sport–Recreation Framework.

The third layer, *digital communities and inclusion*, represents digitally mediated social interactions and community-building practices. Primarily operating at the meso-level, this dimension focuses on digital platforms designed to enhance inclusivity, accessibility, and broader participation within sport and recreation communities. The fourth and final layer, *smart environments and sustainability*, addresses macro-level technology applications within infrastructure and environmental management. This dimension underscores the strategic role of advanced technological solutions in promoting sustainable practices, optimizing resource management, and contributing positively to broader societal well-being. Collectively, these four layers provide a structured analytical framework for understanding the progression of technological mediation within the broader TSR ecosystem, where progress is from micro-level individual performance enhancement, through meso-level immersive and community-oriented interactions, to macro-level societal and environmental outcomes.

Aligning the TSR Framework with Thematic Sections and Chapters

This section illustrates how the conceptual layers of the TSRF correspond to the thematic structure of this book. Specifically, each layer is discussed to demonstrate its alignment with the relevant thematic sections and their corresponding chapters. By mapping these connections, this section aims to clarify how individual chapters collectively contribute to a comprehensive understanding of adoption, application, and implications of technology within sport and recreation contexts. Table 1.1 presents a summary of the layers, including each layer's focus and the corresponding chapters.

Layer 1: Performance Enhancement

Performance enhancement, Layer 1, focuses on the role of technology in optimizing athletic and recreational performance through advanced data analytics, personalized technology applications, and gamification. Chapter 2 discusses the adoption and application of wearable technologies in personal fitness, emphasizing how continuous monitoring and personalized feedback encourage active living and improved athletic performance. Wearable devices, including smartwatches, facilitate precision in goal-setting and achievement by employing visual and auditory cues, thus enhancing user motivation and accountability. Complementing this, Chapter 3 explores the concept of gamification within fitness training, illustrating how game-like elements foster increased engagement and motivation among users. The transformation of gamified fitness from niche gaming culture into mainstream exercise practices demonstrates how interactive technology has reshaped traditional fitness behaviors, potentially broadening access to physical activity. In the same vein,

TABLE 1.1 Aligning the TSR framework with thematic sections and chapters

Layer	*Focus*	*Corresponding Chapters*
Performance enhancement	Sensor-rich, data-driven tools optimizing athletic and recreational performance	Chapters 2–4
Immersive experiences	Virtual and augmented realities and digital extensions reshaping practice and spectator experiences	Chapters 5–9
Digital communities and inclusion	Online networks and adaptive technologies enhancing participation, social connection, and accessibility	Chapters 10–12
Smart environments and sustainability	Technology's role in optimizing venues, event management, and ecological responsibility	Chapters 13–14

Chapter 4 provides a critical examination of sports analytics as a technological intervention for performance enhancement. It situates sports analytics within broader sociopolitical contexts, highlighting its role in systematically tracking, measuring, and assessing athletic performance through data-driven methodologies. While analytics offer objective performance insights, the chapter critically addresses concerns about the marginalization of esthetic aspects of sports, the potential dehumanization of athletes, and ethical challenges around data privacy and surveillance. Collectively, these chapters cover Layer 1 of the TSR framework, demonstrating how technology-driven solutions, ranging from personalized wearables and gamification strategies to sophisticated analytics, serve as critical mechanisms for performance optimization, user engagement, and individualized athlete development.

Layer 2: Immersive Experiences

Immersive experience, Layer 2, examines the role of emerging technologies in creating engaging and interactive environments within sport and recreation contexts. Chapter 5 addresses the expanding influence of AI, discussing how AI-powered applications such as predictive analytics, immersive platforms, and intelligent wearables significantly alter athlete training, injury prevention, and fan engagement. This chapter also underscores ethical and legal challenges, including data privacy and algorithmic bias, emphasizing the necessity of transparent, responsible AI integration to support inclusive and ethical practices. Chapter 6 further elaborates on immersive technologies by specifically examining the utilization of VR and AR. These technologies have evolved significantly and are increasingly integrated into athlete training programs, injury rehabilitation processes, broadcasting enhancements, and spectator experiences. Despite their potential, Chapter 6 highlights barriers to widespread adoption, including high implementation costs and technical complexity.

Chapter 7 extends the discussion into the esports domain, analyzing the environment where digital technology significantly impacts player performance and industry ethics. The rapid growth of esports, accelerated by advancements in digital platforms and online interactions, introduces unique challenges such as player burnout and inconsistent governance structures. Chapter 7 emphasizes the importance of prioritizing athlete well-being through ethical management, career support systems, and wellness initiatives to ensure sustainable growth within esports. Further contributing to the discourse on technological mediation, Chapter 8 explores the digital transformation experienced by traditional sports through advancements in AI, blockchain, immersive technologies, and the emergence of concepts like the metaverse. Lastly, Chapter 9 examines sport-tech startups, focusing on innovations driven by VR, AR, wearable technology, and AI. It highlights emerging trends such as hybrid fitness models and evolving data privacy concerns, illustrating how innovative business strategies and consumer-driven demand reshape the broader fitness and recreation industry landscape. Collectively, these chapters comprehensively illustrate Layer 2 of the TSRF by demonstrating how immersive technologies such as AI, VR, AR, esports platforms, and digital innovations profoundly reshape user experiences, performance paradigms, ethical considerations, and organizational practices across sport and recreation.

Layer 3: Digital Communities and Inclusion

Digital communities and inclusion, Layer 3, addresses the role of digital platforms and adaptive technologies in reshaping social interactions, fostering inclusivity, and enhancing community engagement within sport and recreational contexts. Chapter 10 discusses the evolution of sport communities from localized, face-to-face fan interactions into dynamic digital communities facilitated by contemporary digital platforms. This transformation has significantly expanded fan connectivity, interactivity, and global engagement, redefining traditional notions of sport fandom into a more inclusive and accessible phenomenon that transcends geographic boundaries.

Chapter 11 emphasizes inclusivity through adaptive sports and recreation, highlighting how technological advancements have enhanced performance, participation, and social engagement opportunities for individuals with disabilities. Innovations such as specialized sporting equipment, prosthetics, wearable performance trackers, AI-driven training tools, and smart communication systems have facilitated greater access and improved athlete experiences at both elite and recreational levels. Despite these advances, the chapter underscores persistent challenges, including limited awareness of adaptive sports opportunities, high equipment costs, and uneven access to structured programs, particularly affecting individuals with complex disabilities. Collectively, these chapters underscore Layer 3 of the TSRF by illustrating the transformative impact of digital platforms and adaptive technologies in fostering inclusive, engaging, and globally connected sport and recreational communities.

Layer 4: Smart Environments and Sustainability

Smart environments and sustainability, Layer 4, emphasizes the integration of advanced technologies in managing sport venues, event operations, and the broader environmental impacts associated with sport and recreational activities. Chapter 12 emphasizes that effective adoption of technological innovations not only optimizes event management but also aligns organizational practices with broader sustainability goals, ensuring events remain engaging, equitable, and secure. Complementing this perspective, Chapter 13 explores the critical intersection of technology, sport, and sustainability, underscoring how emerging technological solutions contribute directly to sustainability objectives within sport and recreation contexts. This chapter underscores how technology and innovation are shaping community and outdoor recreation in order to minimize and address environmental impacts. Collectively, these chapters articulate Layer 4 of the TSR Framework by demonstrating how technological advancements enable the creation of smart, sustainable, and efficient sport environments that not only enhance stakeholder experiences but also support overarching environmental and sustainability imperatives.

Summary

This chapter systematically examines the triad of TSR by organizing its interactions within the TSRF. The TSRF offers an integrative perspective that enhances scholarly understanding of technology's pervasive influence across various dimensions of sport and recreation. Each chapter aligns with the TSRF model, providing empirically grounded insights that bridge theoretical and practical domains. This structured approach facilitates a comprehensive understanding and exploration of how technological innovation continues to shape the contemporary landscape of sport and recreation.

References

Abeza, G., & Seguin, B. (2024). International sport marketing. In MacIntosh, E., Bravo, G., and Li, M. (Eds.), *International Sport Management* (3rd ed.). Human Kinetics.

Billings, A. C., Butterworth, M. L., & Turman, P. D. (2017). *Communication and Sport: Surveying the Field*. Sage Publications.

Coakley, J. (2020). *Sports in Society: Issues and Controversies* (13th ed.). McGraw-Hill Education.

Glebova, E., Su, Y., Desbordes, M., & Schut, P. O. (2025). Emerging digital technologies as a game changer in the sport industry. *Frontiers in Sports and Active Living*, *7*, 1605138.

Hurd, A., Anderson, D. M., & Mainieri, T. (2023). *Kraus' Recreation and Leisure in Modern Society* (12th ed.). Jones & Bartlett Learning.

Lee, K. K., & Uihlein, M. J. (2018). Adaptive sports in the rehabilitation of the disabled veterans. *Physical Medicine and Rehabilitation Clinics of North America*, 30(1), 289–299.

Yadav, M. S., & Pavlou, P. A. (2020). Technology-enabled interactions in digital environments: A conceptual foundation for current and future research. *Journal of the Academy of Marketing Science*, *48*, 132–136.

PART I

Technology and Performance Enhancement

2

WEARABLE TECHNOLOGY AND PERSONAL FITNESS

Niveditha Karthikeyan, Susmit S. Gulavani, James Du, and Jeffrey D. James

Introduction

In recent years, the landscape of monitoring and managing personal fitness has transformed with an increasing utilization of wearable technology (Ferguson et al., 2022). During a presentation at Massachusetts Institute of Technology in 1999, Ashton (2009) put forward a term called the Internet of Things (IoT), where he envisaged a fantasy world with inter(net)-connected things using sensors and actuators that can generate real-time data and improve people's quality of daily life. IoT is commonly defined as a network of physical and virtual "things"—such as devices, sensors, or objects—that can communicate, exchange data, and sometimes make decisions using the internet (Li et al., 2015). Within the context of IoT, wearable technology was presented as the next generation of market demands following smartphones (Ometov et al., 2021). Bower and Sturman (2015) define wearable technologies as "digital devices that incorporate wireless connectivity for the purposes of seamlessly accessing, interacting with and exchanging contextually relevant information" (p. 344). They include body-worn devices like smartwatches, fitness trackers, smart glasses, and smart apparel that offer unprecedented access to real-time health data as well as other personalized insights related to fitness, activity, and daily routines (Kalantari, 2017; Pizzo et al., 2020). These wearable devices can be paired with smartphone applications to enhance user engagement and motivation by providing features for goal setting, performance measurement and tracking, and providing feedback on various types of physical activities. Further, wearable devices transcend the elementary-level data (e.g., steps walked or calories burned tracked by Fitbit Inspire 3) and provide insights (e.g., Apple Series 10 wristwatch) into heart rate measurement and energy expenditure, underscoring their potential as tools for health monitoring.

DOI: 10.4324/9781003588405-3

While recognizing the benefits of wearable technology in promoting active living among individuals, the adoption of wearable technology has certain challenges. For example, the data collected by wearable devices often raises concerns regarding privacy, security, and informed consent (Sivakumar et al., 2024). Also, the potential misuse of sensitive and personal health data could carry repercussions for the concerned individuals (Jessop & Baker, 2019). Addressing these challenges requires ongoing innovation in device design, data analytics, and user experience, along with collaboration between researchers, developers, healthcare professionals, and regulatory bodies (Sivakumar et al., 2024). It can be argued that as wearable technology continues to evolve, these devices hold immense promise for transforming personal fitness and fostering a culture of proactive health and wellness management. This chapter explores the evolution, applications, and implications of wearable technology in personal fitness, focusing on its role in enhancing athletic performance and promoting health and wellness. We cover key developments, concepts, ethics, trends, and an elite sports case study, offering insights for researchers, professionals, and general readers.

Background

Wearable technologies for tracking fitness and performance have come a long way from simple mechanical versions of pedometers (which count the number of steps a person takes) to sophisticated sensor-driven devices for health and performance monitoring. As far back as the early 2000s, the integration of sensors and GPS technology began reshaping how fitness was tracked, both for athletes and everyday users. Figure 2.1 presents a timeline of key innovations that illustrate the evolution of wearable technology in fitness and sport.

Building on the evolution shown in the timeline, the 2000s marked the beginning of widespread adoption of wearable devices, with Fitbit's launch in 2007 signaling a pivotal shift in consumer fitness tracking (Ometov et al., 2021). This era also saw the integration of GPS and sensors, enhancing functionality and accessibility. Notably, Nike's 2006 collaboration with Apple introduced a shoe sensor that allowed runners to track performance, paving way for future innovations (Apple Inc., 2006). The 2010s brought game-changing devices, including Fitbit's wireless tracker (in 2009) (Miller, 2008), Jawbone Up (in 2011) (PR Newswire, 2011), and the Apple Watch (in 2015) (Apple Inc., 2014), which introduced features like heart rate monitoring, sleep tracking, and electrocardiography (ECG, which measures the heart's electrical activity). These devices enabled users to monitor both fitness and key health indicators as part of daily routines (Niknejad et al., 2020). The release of the Whoop (2015) fitness band further pushed the boundaries of performance tracking by offering advanced insights into recovery, strain, and sleep quality, thereby

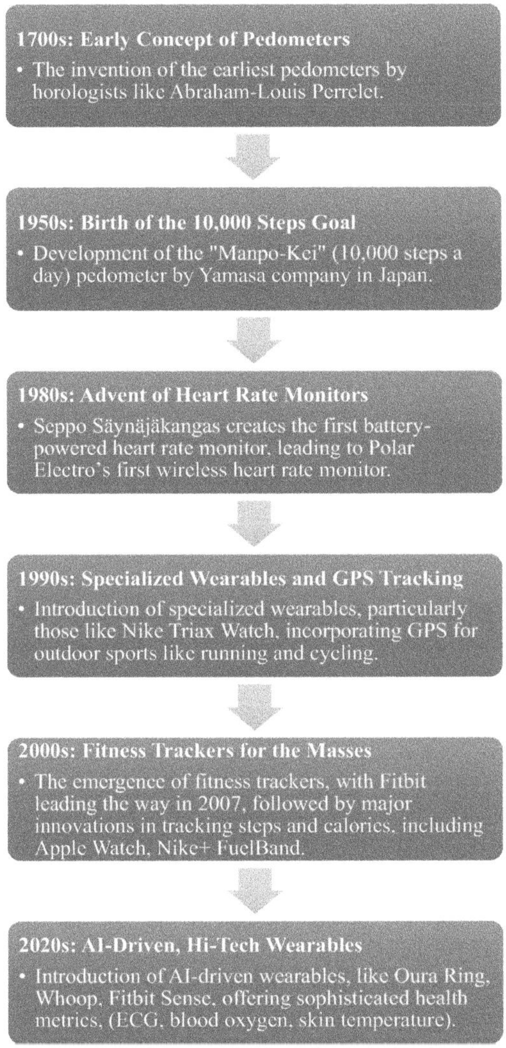

FIGURE 2.1 Evolution of wearable technology in fitness and performance tracking.

emphasizing the importance of recovery in athletic performance and providing personalized recommendations, a step beyond the more generic features of earlier wearable technologies.

In the 2020s, wearable technologies evolved to incorporate more advanced features driven by artificial intelligence (AI) and machine learning. Devices like the Oura Ring and Apple Watch 7 (in 2021) introduced critical health-monitoring capabilities such as blood oxygen level tracking and ECG readings, blurring the lines between fitness trackers and medical devices, albeit with a lesser degree of sophistication. By 2022, wearable technologies like the Whoop 4.0 fitness band began offering more personalized data, with metrics like skin temperature and real-time AI-driven coaching, enhancing the ability to optimize performance and recovery. As wearable technologies continue to advance, even more sophisticated capabilities, including AI-powered predictive analytics, ultraviolet (UV) exposure tracking, and hydration monitoring, make it possible to tailor user actions to different recommendations. These innovations promise to further personalize fitness and health insights, providing real-time data and tailored recommendations for optimizing both wellness and performance.

Topical Exploration and Key Concepts

Definition and Taxonomy

Wearable technology refers to digital devices that use wireless connectivity to provide users with real-time access to and interaction with information (Bower & Sturman, 2015). These devices are designed to be worn on the body—as wristbands, rings, or patches—seamlessly integrating into everyday life to provide valuable insights about bodily parameters. In sports, wearable technology can be understood through a three-tiered framework: sensors, processing, and communication. The sensor layer captures biometric data through physiological sensors like ECG, electromyography (EMG), and electroencephalography (EEG), as well as inertial measurement units (IMUs) that track motion for applications like swimming and posture analysis (Aroganam et al., 2019). The processing layer uses microcontrollers to manage tasks like heart rate and step counting, while the communication layer relies on Bluetooth and Wi-Fi connectivity for data transmission and cloud computing for advanced processing. Communication is an important aspect as it provides data that can be used for gleaning actionable insights, such as alerts for abnormal heart rates or reminders to increase daily activity. Wearable technologies are primarily made up of accelerometers, gyroscopes, and magnetometers, which analyze movements and the forces, providing insights into kinematics and kinetics to optimize sports performance. These technologies are compact, discreet, and provide personalized data, enabling remote monitoring through a mere network integration.

Athletes, coaches, and other stakeholders benefit from the wealth of information that wearable technologies provide, including vital signs monitoring and performance tracking. Unlike smartphones or tablets, wearable technologies offer features like monitoring, biofeedback, and physiological sensing. They are used for health monitoring, chronic disease management, performance tracking, and even interactive gaming. The rise of wearable technology aligns with the "Quantified Self" movement (Lupton, 2016), which emphasizes the use of devices to generate measurable data about daily life, empowering individuals to track their health, fitness, and athletic performance in real time. Building on the diverse applications of wearable technologies, it is important to understand the various types and classifications of these devices to better grasp their unique functionalities and impact.

Taxonomy and Types

Wearable technologies can be classified based on functionality, form, and body placement, with the most common classification based on their application. Devices that are worn or carried can serve similar functions but vary greatly in design, technology, and placement on the body. For example, a smartwatch and a body-mounted sensor both track activity, but one is worn on the wrist while the other may be clipped onto clothing. The modern wearable technologies market is vast, leading to a variety of device classification approaches. Table 2.1 provides a broad classification of wearable types based on their application—while it is not exhaustive due to the wide variety of wearable technologies available and their continuous growth, it highlights the most popular ones at present (in 2025). As the wearable market continues to expand, understanding how users adopt and are motivated to use these devices becomes crucial.

TABLE 2.1 Types of wearable technologies

Type of Wearable Technologies	*Description*	*Product Examples*
Smart watches	Wearable technologies that combine traditional watch functions with smart features allow users to perform tasks like tracking fitness, receiving notifications, and interacting with apps, all from their wrist.	Apple Watch, Samsung Galaxy Watch, Garmin Vívomove.
Smart jewelry	Devices like rings, wristbands, and watches that sync with smartphone apps.	Oura Ring, Joule Earring Backings.
Fitness trackers	Wearable technologies that are designed to track physical activity, steps, heart rate, and other vital signs, usually connected to apps for detailed data analysis.	Fitbit Charge, Garmin Vivosmart, Whoop Strap.

(Continued)

TABLE 2.1 (Continued)

Type of Wearable Technologies	*Description*	*Product Examples*
Smart clothing	Clothing that is integrated with technology for fitness tracking, health monitoring, and other functionalities.	Hexoskin, Athos, Under Armour Smart Shirt.
Body-mounted sensors	Wearable technologies that are typically placed on the body (e.g., on the skin or attached to clothing) to continuously monitor and collect biological data, often for specialized health or medical purposes.	Biostrap, Tempdrop, Catapult Tracker, STATSport GPS Tracker.
Augmented Reality (AR) headsets	Devices that blend digital content with the real world, overlaying virtual elements onto the user's physical environment.	Microsoft HoloLens, Magic Leap One, Nreal Light.
Virtual Reality (VR) headsets	Wearable technologies that immerse the user fully in a simulated digital environment, replacing the real world with virtual surroundings.	Oculus Quest 2, PlayStation VR, HTC Vive.
Smart tattoos	Temporary tattoos embedded with sensors to monitor health metrics like hydration or glucose levels in real time.	Prinker, BioStampRC, SkinPatch.

Smartwatches are the most popular wearables, combining traditional timekeeping with fitness tracking, health monitoring, and app interaction (e.g., Apple Watch). Smart jewelry integrates technology into accessories like rings and wristbands, offering discreet ways to track fitness and health metrics without the bulk of traditional wearables (e.g., Oura Ring). Fitness trackers focus on monitoring activities like step counting, calorie burn, and heart rate, syncing with apps to offer comprehensive health insights (e.g., Fitbit Charge). Smart clothing incorporates sensors into textiles for real-time performance analytics, such as heart rate and muscle activity (e.g., Hexoskin). Body-mounted sensors are typically worn on the body or attached to clothing, providing precise biometric data for athletes and medical professionals (e.g., Biostrap). Lastly, AR and VR headsets create immersive environments for fitness training and health monitoring (e.g., Oculus Quest 2). Smart tattoos offer a noninvasive, discreet way to monitor health metrics like glucose levels in real time (e.g., BioStampRC). Each type offers unique benefits for fitness and health enhancement.

Role and Benefits in Athletic Performance

The growing use of wearable technology in sports offers unobtrusive measurement systems that provide real-time data on various performance and injury risk indicators, such as range of motion, accelerations, and impacts. This objective data enhances coaching by enabling more informed decisions, but to be effective, the information must be presented in an easily interpretable format. For athletes, wearable technologies provide auditory or vibratory feedback that helps them focus on their movements and environment, enabling biomechanical adjustments without the need for a coach. A major advantage of wearable technologies is their ability to monitor athletes in real-world environments, delivering real-time feedback (Adesida et al., 2019). Additionally, their small, lightweight, wireless, and unobtrusive design allows athletes to move freely during activities. This enables athletes to be monitored outside laboratory settings and within their natural training environments. Wearable technologies in sport are primarily used for two reasons: (a) injury prevention and rehabilitation and (b) performance optimization.

Injury Prevention and Rehabilitation

Wearable technologies enable early detection of abnormal movement patterns or physiological responses, helping identify injury risks before they worsen. These technologies support evidence-based decision-making, facilitating personalized interventions. For example, in American football, wearable sensors embedded in helmets and mouthguards can monitor concussions by detecting linear and angular head accelerations during impacts. Siegmund et al. (2015) demonstrated that such systems could detect over 95% of impacts, providing invaluable data for proactive interventions. By monitoring athlete biomechanics, wearable technologies allow for targeted interventions like corrective exercises or training adjustments. Machine learning can further enhance injury prevention by analyzing large datasets to create personalized programs (Kiernan et al., 2018). Wearable technologies play a key role in minimizing sports-related injuries by providing real-time data to monitor and prevent potential risks.

In rehabilitation, wearable technologies track an athlete's progress, providing real-time feedback on movement and muscle activation, aiding healthcare professionals in personalizing recovery protocols. They also allow clinicians to tailor exercises based on specific deficits or challenges, enhancing the rehabilitation process. For instance, wearable sensors have improved functional outcomes and reduced recovery times for young athletes recovering from ACL injuries (Vijayan et al., 2021). Additionally, accelerometer data has been used to predict running injuries in National Collegiate Athletics Association (NCAA) athletes, helping prevent overuse injuries (Kiernan et al., 2018). Wearables thus support a data-driven approach to minimizing injury risks and improving rehabilitation.

Performance Optimization

Optimizing performance involves quantifying movements to refine the technique of athletes (Chidambaram et al., 2022). For example, FORM Swim Goggles provides real-time feedback on stroke efficiency during swimming, and devices like Stryd pods help runners analyze their mechanics. Wearable technologies can further enable biomechanical feedback by tracking functional movements, acceleration, and force, offering insights into movement patterns (Adesida et al., 2019). Wearable technologies also monitor physiological indicators like heart rate, blood pressure, and oxygen saturation to enhance performance (Wisbey et al., 2010). These tools help coaches refine performance analysis and training methods.

Wearable technologies offer real-time monitoring of performance and physiological metrics, aiding in player health management and performance optimization through data analysis of workload and physiological markers (Adesida et al., 2019). According to Osborne (2017), wearable data helps monitor player load, optimize training/rest, and prevent injury. High training volumes can drive success, but excessive intensity risks fatigue and overuse injuries (Penttä & Hassmén, 1998). Therefore, performance is aimed to be "optimized" rather than "maximized." Monitoring both training and competition loads is crucial for making informed decisions and minimizing injury risks (Osborne, 2017). Real-time analysis allows for identification of performance patterns and tailoring of training programs to enhance athletic performance (Chidambaram et al., 2022). Wearable technologies, while enhancing performance and injury prevention, also offer significant health and wellness benefits in recreational contexts, supporting a wide range of users in their everyday physical activities.

Health and Wellness Benefits

The proliferation of wearable technology has revolutionized personal fitness, making physical activity (PA) more motivating, personalized, and goal-oriented. Wearable technologies like smartwatches and fitness trackers record real-time data on heart rate, steps walked, calories burned, and sleep patterns (Ferreira et al., 2021), encouraging individuals to shift from passive leisure to more active, health-conscious pursuits (Du et al., 2021). These devices are particularly helpful in combating sedentary lifestyles, as they foster active participation in recreational activities. Furthermore, they enhance health monitoring and chronic disease prevention efforts, while incorporating gamification, which uses game-like elements such as earning badges for fitness goals (Richter et al., 2015), and social engagement to make wellness more enjoyable and competitive.

Enhancing Physical Wellness

PA levels are declining globally, increasing chronic disease risks (Ferguson et al., 2022). Wearable technology can promote behavior change by supporting self-monitoring, goal setting, and self-regulation (Stragier et al., 2018). Wearable technologies, such as fitness trackers, smartwatches, and pedometers, are affordable (starting at $30), user-friendly, and promote active leisure by reducing barriers to exercise (Maher et al., 2017). Through a systematic review, Ferguson et al. (2022) found that wearable technology positively impacts daily step counts, PA, and energy expenditure, with moderate effects on moderate-to-vigorous PA. Wearable technologies also sustain PA by providing consistent feedback and motivation (Bort-Roig et al., 2014).

Trials using devices like the Fitbit One have shown increased PA, though results vary with intervention duration (Cadmus-Bertram et al., 2015). In older adults, wearable technology has improved physical and functional health (McMahon et al., 2017), with pedometers increasing PA by an average of 2,000 steps per day (Kang et al., 2009). The benefits of wearable technologies include improved physiological outcomes, such as reduced BMI and blood pressure (Bravata et al., 2007), and enhanced aerobic capacity (Qiu et al., 2018). PA has also been linked with improved mental health, including reductions in depression and anxiety (Kandola et al., 2019; Stubbs et al., 2017). The use of wearable technologies in clinical trials continues to rise, demonstrating their potential in improving health outcomes, such as weight loss and increased PA in diverse populations (Cadmus-Bertman et al., 2015). Wearable technologies are therefore a valuable tool for promoting PA, improving health outcomes, and supporting behavior change across diverse populations, with their effectiveness continually being explored in both general wellness and clinical settings.

Social Engagement

In addition to tracking PA, wearable technologies leverage gamification and social engagement to boost motivation. These devices incorporate challenges, rewards, and progress tracking, creating a sense of achievement and fostering a competitive yet supportive environment. Social elements like community challenges and leaderboards enhance accountability, motivating users to engage in active leisure (Baert et al., 2011). Group-based activities fulfill social needs, with social interaction boosting intrinsic motivation (Richter et al., 2015). However, when rewards become extrinsically motivating, such as through points or monetary prizes, they can detract from intrinsic interest in PA.

Wearable technologies also leverage rewards-based gamification, reinforcing positive behavior through social comparison and peer support (Richter et al., 2015). This approach creates a dynamic, engaging experience, encouraging sustained PA and improving long-term health outcomes outside of competitive environments.

Privacy, Security, and Ethical Considerations

Wearable technologies collect vast amounts of personal data, often accessed by third parties without user consent, raising ethical concerns about privacy, security, and informed consent (Sivakumar et al., 2024). The benefits of wearable technology in personal fitness and health must be carefully weighed against their potential risks, including data breaches and leaks, which can result in serious privacy violations (Niknejad et al., 2020). Security issues, such as unauthorized access or data loss, also jeopardize the confidentiality and integrity of collected data. In sports, the increasing use of wearable technologies has created challenges in distinguishing between personal health data and performance-related data, with unclear ownership of and unrestricted access to such information. Scholars have raised concerns about the potential misuse of athletes' physiological data, highlighting risks such as data privacy issues and possible career repercussions for athletes (Jessop & Baker, 2019; Karkazis & Fishman, 2017). Athlete data often includes sensitive health and performance information, including medical details (Garlewicz, 2020). However, organizations compiling this data may claim ownership, reducing an athlete's agency over their physical body. As the use of this data extends beyond sport, the ethical dilemma intensifies, outpacing current legislative measures and creating broader concerns around data ownership, intellectual property, privacy, and security. These issues expose athletes to significant risks related to privacy and career security (Karkazis & Fishman, 2017), stressing the need for robust ethical frameworks governing data collection, use, and protection. Ensuring users are informed about data collection practices and have control over their data remains a significant challenge for developers and regulators (Sivakumar et al., 2024).

As wearable data usage expands beyond sports, ethical concerns grow, especially around data analysis methods. Data mining techniques, such as clustering and classification, are commonly used in other industries such as marketing and banking to create customer profiles (Sivakumar et al., 2024). However, these profiles can lead to stigmatization and discrimination if sensitive information, such as health data, is shared without consent. Lupton (2016) highlighted privacy concerns among users of health and smart wearable technologies, identifying key factors influencing users' privacy perceptions, including data sensitivity and transparency. Users often feel pressured into accepting data collection terms, and the lack of transparency exacerbates these concerns. Users prefer not to share their data on public forums, emphasizing a desire for greater control over personal information. These findings underscore the importance of stronger privacy protections and clearer communication about data usage, promoting ethical frameworks in wearable technology to safeguard user privacy and build trust.

Trends and Current Issues

AI in Wearable Technology

The use of AI in wearable technology has revolutionized how individuals manage their fitness, health, and overall well-being. By analyzing personal data, AI-powered devices can generate customized fitness programs, dietary recommendations, and stress-relief strategies (Nahavandi et al., 2022). These technologies provide real-time feedback on activity levels, sleep quality, and health indicators, enabling users to make informed choices (Ramanujam et al., 2021). Furthermore, AI-driven algorithms can proactively identify potential health risks, allowing individuals to take preventive steps to maintain their well-being. Additionally, AI enhances the user experience by adapting to behaviors over time (Nahavandi et al., 2022).

AI-powered wearable technology serves multiple purposes, including fitness tracking, health assessment, and stress management. Smart fitness trackers equipped with AI can recognize different exercises and provide detailed performance analysis (Nahavandi et al., 2022). Similarly, AI-enhanced smartwatches monitor vital health metrics, detect irregularities, and offer tailored recommendations (Ramanujam et al., 2021). Beyond health monitoring, AI-powered wearable technologies play a crucial role in enhancing personal fitness. These devices can generate individualized workout routines based on fitness goals, current ability, and available time (Nahavandi et al., 2022). They also deliver real-time coaching, helping users refine their technique and maximize workout efficiency. By tracking progress over time, AI wearables serve as a source of motivation, encouraging users to stay committed to their fitness journey.

Sustainable Wearable Technology

Sustainable wearable technologies aim to improve the quality of life by adopting human-centric applications and services that go beyond simply measuring vital body parameters. Sustainable wearable technologies focus on continuously tracking and monitoring outcomes of various environmental events that change over time, contributing to public interest and safety (Lee et al., 2016). For example, wearable technologies can monitor air quality in real-time, providing users with actionable data to reduce exposure to pollutants, or track water usage to promote conservation efforts. Energy efficiency is a key feature of sustainable wearables, often incorporating energy-harvesting technologies like solar power or kinetic energy to reduce reliance on traditional batteries (Chong et al., 2019). This not only enhances the longevity of the devices but also aligns with global efforts to reduce carbon footprints.

Additionally, sustainable design principles emphasize the reuse and recycling of wearable components, promoting a circular economy to minimize electronic waste and extend product life cycles (Gurova et al., 2020). For instance, modular designs allow users to replace or upgrade individual components rather than discard entire devices, reducing environmental impact. By integrating both human life improvements and sustainability values, sustainable wearable technologies offer a social impact that benefits society at large, fostering a healthier relationship between technology, individuals, and the environment.

Virtual Reality (VR), Augmented Reality (AR), and Mixed Reality (MR) in Wearable Technology

Wearable technologies have become increasingly integrated with VR, AR, and MR, creating immersive and interactive experiences that bridge the physical and digital worlds. VR headsets like the Meta Quest and HTC Vive Pro enable athletes to train in 360-degree environments, offering opportunities for skill development, tactical exercises, and rehabilitation. For example, VR is now used to simulate race courses for cyclists or runners, helping athletes to train in virtual environments that replicate outdoor conditions (Akbaş et al., 2019). This technology is also widely used in injury recovery, where users can engage in guided rehabilitation exercises through VR simulations, promoting movement and recovery in a safe, controlled space.

AR wearables, such as the Solos Smart Sport Sunglasses, provide real-time feedback during workouts, displaying performance metrics such as heart rate, pace, and posture correction (Neges et al., 2018). These devices help improve athletic performance by offering immediate, contextual insights that enhance exercise routines. For example, AR technology can overlay data onto the user's field of view during a run, offering real-time performance tracking, which supports goal setting and motivates continued improvement. MR wearables, such as the Magic Leap headset, blend virtual and physical environments to enhance fitness training and rehabilitation by enabling immersive, personalized workouts and real-time performance tracking. The integration of biometric sensors like heart rate monitors allows for adaptive experiences, adjusting workouts based on real-time physiological data (Adesida et al., 2019). As MR, AR, and VR technologies evolve, they are poised to transform fitness and wellness, offering more interactive and personalized experiences.

Practice in Action

To explore the impact of wearable technology on athletic performance, we gathered insights from professionals in cricket and field hockey, including coaches, trainers, and data analysts. These professionals discussed how wearable technology has transformed fitness monitoring, performance tracking, injury prevention, and workload management.

Integrating Wearable Technologies into Training

The implementation of wearable technologies in elite sports was primarily driven by the need for performance monitoring, injury prevention, and recovery management. Wearable technologies were used to measure Key Performance Indicators (KPI) like running distance, acceleration, and heart rate variability. One coach explained, "By monitoring the load and performance of athletes, we were able to decrease injury rates and make decisions on load/recovery." In this example, athletes underwent structured onboarding, learning about the technology's benefits and receiving continuous feedback to stay engaged.

Performance Enhancements and Adaptations

Wearable technologies led to noticeable improvements in performance and injury reduction. One coach shared, "We were able to have fewer injuries and better performance," while another highlighted, "Data helped adjust a veteran player's workload, leading to a significant boost in performance in just a week." They also enhanced players' conditioning and recovery by tracking heart rate, workload, and movement intensity. A trainer leveraged insights from wearable technology to compare training session data with in-game performance, identifying discrepancies in heart rate and velocity, particularly among younger athletes, enabling targeted interventions to enhance match preparedness.

Challenges and Lessons Learned

A major challenge was the overwhelming amount of data generated. One analyst noted, "The biggest challenge was the time needed to get the data sorted from the GPS. It was a lot of work for the strength and conditioning coach." The onboarding process was key in overcoming resistance: "Once the onboarding sessions were done right, the rest took care of itself," said one trainer. However, issues like privacy concerns, data overload, and cost arose, with cheaper wearable technologies lacking sufficient storage and requiring frequent syncing. Comfort also varied among players, but proper onboarding helped boost engagement as athletes saw the performance benefits.

Key Takeaways

1. Gradual Integration: Start with a few critical metrics and expand over time to prevent overwhelming coaching staff and players.
2. Real-Time Feedback: Providing immediate insights enhances player motivation and performance engagement.
3. Investment in Quality Wearables: Cutting costs on wearable technology can limit data accuracy and long-term benefits.
4. Athlete Engagement: Encouraging competition and self-improvement through data-driven insights increases adoption rates.
5. Customization Matters: Personalizing wearable technologies and data analysis to suit player needs ensures better compliance and effectiveness.

By learning from these professionals' experiences, users can optimize their approach to wearable technology, improving performance outcomes while overcoming common obstacles in implementation.

Summary

This chapter explores the evolution, applications, and implications of wearable technology in personal fitness, with a focus on its role in enhancing athletic performance, promoting health and wellness, and the integration of AI for personalized insights. The adoption of wearable technology has empowered individuals to monitor their personal fitness continuously, fostering a culture of accountability (Pizzo et al., 2020). Devices like smartwatches use visual and auditory cues, such as flashing lights or sounds, to alert users when they fall short of activity goals. By providing reminders to move, celebrating milestones, and offering actionable feedback, these devices keep users motivated and engaged in their fitness journeys. Wearable technologies have shifted personal fitness from a one-size-fits-all approach to a highly personalized experience, allowing users to set specific goals and track their progress with precision. Moreover, they have made advanced health insights accessible to the general public, not just athletes or fitness professionals (Ferguson et al., 2022).

As the use of wearable technology grows, advancements in AI can further enhance these devices by analyzing complex fitness and health data to deliver personalized recommendations (Chidambaram et al., 2022). Enhanced personalization boosts user engagement and motivation for an active lifestyle. While wearable technologies already encourage a proactive approach to personal fitness and active living, integrating AI can revolutionize the landscape by providing real-time feedback and coaching. Advanced AI algorithms enhance the accuracy of monitoring vital body parameters and optimize athletic performance, with the potential to deliver more sophisticated, proactive, and user-centric solutions as AI evolves, bridging the gap between fitness, health, and technology.

Key Resources

- **Whoop:** https://www.whoop.com
- **Fitbit:** https://www.fitbit.com
- **Strava's Instagram:** https://www.instagram.com/strava
- **MagicLeap:** https://www.magicleap.com
- **Hexoskin:** https://www.hexoskin.com

References

Adesida, Y., Papi, E., & McGregor, A. H. (2019). Exploring the role of wearable technology in sport kinematics and kinetics: A systematic review. *Sensors*, *19*(7), 1597. https://doi.org/10.3390/s19071597

Akbaş, A., Marszałek, W., Kamieniarz, A., Polechoński, J., Słomka, K. J., & Juras, G. (2019). Application of virtual reality in competitive athletes–A review. *Journal of Human Kinetics*, *69*, 5.

Apple Inc. (2006, May 23). *Nike and Apple Team Up to Launch Nike+iPod* [Press release]. https://www.apple.com/newsroom/2006/05/23Nike-and-Apple-Team-Up-to-Launch-Nike-iPod/

Apple Inc. (2014, September 9). Apple unveils Apple Watch—Apple's most personal device ever [Press release]. *Apple Newsroom*. https://www.apple.com/newsroom/2014/09/09Apple-Unveils-Apple-Watch-Apples-Most-Personal-Device-Ever/

Aroganam, G., Manivannan, N., & Harrison, D. (2019). Review on wearable technology sensors used in consumer sport applications. *Sensors*, *19*(9), 1983. https://doi.org/10.3390/s19091983

Ashton, K. (2009). That 'internet of things' thing. *RFID Journal*, *22*(7), 97–114.

Baert, V., Gorus, E., Mets, T., Geerts, C., & Bautmans, I. (2011). Motivators and barriers for physical activity in the oldest old: A systematic review. *Ageing Research Reviews*, *10*(4), 464–474.

Bort-Roig, J., Gilson, N. D., Puig-Ribera, A., Contreras, R. S., & Trost, S. G. (2014). Measuring and influencing physical activity with smartphone technology: A systematic review. *Sports Medicine*, *44*, 671–686.

Bower, M., & Sturman, D. (2015). What are the educational affordances of wearable technologies? *Computers & Education*, *88*, 343–353. https://doi.org/10.1016/j.compedu.2015.07.013

Bravata, D. M., Smith-Spangler, C., Sundaram, V., Gienger, A. L., Lin, N., Lewis, R., Stave, C. D., Olkin, I., & Sirard, J. R. (2007). Using pedometers to increase physical activity and improve health. *JAMA*, *298*(19), 2296. https://doi.org/10.1001/jama.298.19.2296

Cadmus-Bertram, L. A., Marcus, B. H., Patterson, R. E., Parker, B. A., & Morey, B. L. (2015). Randomized trial of a Fitbit-based physical activity intervention for women. *American Journal of Preventive Medicine*, *49*(3), 414–418.

Chidambaram, S., Maheswaran, Y., Patel, K., Sounderajah, V., Hashimoto, D. A., Seastedt, K. P., McGregor, A. H., Markar, S. R., & Darzi, A. (2022). Using artificial intelligence-enhanced sensing and wearable technology in sports medicine and performance optimisation. *Sensors*, *22*(18), 6920. https://doi.org/10.3390/s22186920

Chong, Y. W., Ismail, W., Ko, K., & Lee, C. Y. (2019). Energy harvesting for wearable devices: A review. *IEEE Sensors Journal*, *19*(20), 9047–9062.

Du, J., Floyd, C., Kim, A. C., Baker, B. J., Sato, M., James, J. D., & Funk, D. C. (2021). To be or not to be: Negotiating leisure constraints with technology and data analytics amid the COVID-19 pandemic. *Leisure Studies*, *40*(4), 561–574. https://doi.org/10.1080/02614367.2020.1862284

Ferguson, T., Olds, T., Curtis, R., Blake, H., Crozier, A. J., Dankiw, K., Dumuid, D., Kasai, D., O'Connor, E., Virgara, R., & Maher, C. (2022). Effectiveness of wearable activity trackers to increase physical activity and improve health: A systematic review of systematic reviews and meta-analyses. *The Lancet Digital Health*, *4*(8), e615–e626. https://doi.org/10.1016/s2589-7500(22)00111-x

Ferreira, J. J., Fernandes, C. I., Rammal, H. G., & Veiga, P. M. (2021). Wearable technology and consumer interaction: A systematic review and research agenda. *Computers in Human Behavior*, *118*, 106710. https://doi.org/10.1016/j.chb.2021.106710

Garlewicz, A. (2020). Athlete biometric data in soccer: Athlete protection or athlete exploitation? *DePaul Journal of Sports Law*, *16*, ii.

Gurova, O., Merritt, T. R., Papachristos, E., & Vaajakari, J. (2020). Sustainable solutions for wearable technologies: Mapping the product development life cycle. *Sustainability*, *12*(20), 8444.

Jessop, A., & Baker, T. (2019). Big data bust: Evaluating the risks of tracking NCAA athletes' biometric data. *Texas Review of Entertainment & Sports Law*, *20*, 81.

Kalantari, M. (2017). Consumers' adoption of wearable technologies: Literature review, synthesis, and future research agenda. *International Journal of Technology Marketing*, *12*(3), 274. https://doi.org/10.1504/ijtmkt.2017.089665

Kandola, A., Ashdown-Franks, G., Hendrikse, J., Sabiston, C. M., & Stubbs, B. (2019). Physical activity and depression: Towards understanding the antidepressant mechanisms of physical activity. *Neuroscience & Biobehavioral Reviews*, *107*, 525–539.

Kang, M., Marshall, S. J., Barreira, T. V., & Lee, J. O. (2009). Effect of pedometer-based physical activity interventions: A meta-analysis. *Research Quarterly for Exercise and Sport*, *80*(3), 648–655.

Karkazis, K., & Fishman, J. R. (2017). Tracking US professional athletes: The ethics of biometric technologies. *The American Journal of Bioethics*, *17*(1), 45–60.

Henttä, G., & Hassmén, P. (1998). Overtraining and recovery: A conceptual model. *Sports Medicine*, *26*, 1–16.

Kiernan, D., Hawkins, D. A., Manoukian, M. A., McKallip, M., Oelsner, L., Caskey, C. F., & Coolbaugh, C. L. (2018). Accelerometer-based prediction of running injury in National Collegiate Athletic Association track athletes. *Journal of Biomechanics*, *73*, 201–209.

Lee, J., Kim, D., Ryoo, H. Y., & Shin, B. S. (2016). Sustainable wearables: Wearable technology for enhancing the quality of human life. *Sustainability*, *8*(5), 466.

Li, S., Xu, L. D., & Zhao, S. (2015). The internet of things: A survey. *Information Systems Frontiers*, *17*, 243–259.

Lupton, D. (2016). The diverse domains of quantified selves: Self-tracking modes and dataveillance. *Economy and Society*, *45*(1), 101–122. https://doi.org/10.1080/03085147.2016.1143726

Maher, C., Ryan, J., Ambrosi, C., & Edney, S. (2017). Users' experiences of wearable activity trackers: A cross-sectional study. *BMC Public Health*, *17*, 1–8.

McMahon, S. K., Lewis, B., Oakes, J. M., Wyman, J. F., Guan, W., & Rothman, A. J. (2017). Assessing the effects of interpersonal and intrapersonal behavior change strategies on physical activity in older adults: A factorial experiment. *Annals of Behavioral Medicine*, *51*(3), 376–390.

Miller, C. C. (2008, September 9). Fitness gadget aims to help users lose weight. *Bits Blog*. https://archive.nytimes.com/bits.blogs.nytimes.com/2008/09/09/fitness-gadget-aims-to-help-users-lose-weight/

Nahavandi, D., Alizadehsani, R., Khosravi, A., & Acharya, U. R. (2022). Application of artificial intelligence in wearable devices: Opportunities and challenges. *Computer Methods and Programs in Biomedicine*, *213*, 106541.

Neges, M., Adwernat, S., & Abramovici, M. (2018). Augmented virtuality for maintenance training simulation under various stress conditions. *Procedia Manufacturing*, *19*, 171–178. https://doi.org/10.1016/j.promfg.2018.01.024

Niknejad, N., Ismail, W. B., Mardani, A., Liao, H., & Ghani, I. (2020). A comprehensive overview of smart wearables: The state of the art literature, recent advances, and future challenges. *Engineering Applications of Artificial Intelligence*, *90*, 103529. https://doi.org/10.1016/j.engappai.2020.103529

Ometov, A., Shubina, V., Klus, L., Skibińska, J., Saafi, S., Pascacio, P., Flueratoru, L., Gaibor, D. Q., Chukhno, N., Chukhno, O., Ali, A., Channa, A., Svertoka, E., Qaim, W. B., Casanova-Marqués, R., Holcer, S., Torres-Sospedra, J., Casteleyn, S., Ruggeri, G., … Lohan, E. S. (2021). A survey on wearable technology: History, state-of-the-art and current challenges. *Computer Networks*, *193*, 108074. https://doi.org/10.1016/j.comnet.2021.108074

Osborne, B. (2017). Legal and ethical implications of athletes' biometric data collection in professional sport. *Marquette Sports Law Review*, *28*, 37.

Pizzo, A. D., Baker, B. J., Jones, G. J., & Funk, D. C. (2020). Sport experience design: Wearable fitness technology in the health and fitness industry. *Journal of Sport Management*, *35*(2), 130–143.

PR Newswire. (2011, November 3). *UP™ by Jawbone® with MotionX® Technology Empowers You to Live a Healthier Life* [Press release]. https://www.prnewswire.com/news-releases/up-by-jawbone-with-motionx-technology-empowers-you-to-live-a-healthier-life-133148048.html

Qiu, S., Cai, X., Wang, X., He, C., Zuegel, M., Steinacker, J. M., & Schumann, U. (2018). Using step counters to promote physical activity and exercise capacity in patients with chronic obstructive pulmonary disease: A meta-analysis. *Therapeutic Advances in Respiratory Disease*, *12*, 1753466618787386.

Ramanujam, E., Perumal, T., & Padmavathi, S. (2021). Human activity recognition with smartphone and wearable sensors using deep learning techniques: A review. *IEEE Sensors Journal*, *21*(12), 13029–13040.

Richter, G., Raban, D. R., & Rafaeli, S. (2015). *Studying Gamification: The Effect of Rewards and Incentives on Motivation* (pp. 21–46). Springer International Publishing.

Siegmund, G. P., Guskiewicz, K. M., Marshall, S. W., DeMarco, A. L., & Bonin, S. J. (2015). Laboratory validation of two wearable sensor systems for measuring head impact severity in football players. *Annals of Biomedical Engineering*, *44*(4), 1257–1274. https://doi.org/10.1007/s10439-015-1420-6

Sivakumar, C. L. V., Mone, V., & Abdumukhtor, R. (2024). Addressing privacy concerns with wearable health monitoring technology. *Wiley Interdisciplinary Reviews: Data Mining and Knowledge Discovery*, *14*(3), e1535.

Stragier, J., Vanden Abeele, M., & De Marez, L. (2018). Recreational athletes' running motivations as predictors of their use of online fitness community features. *Behaviour & Information Technology*, *37*(8), 815–827.

Stubbs, B., Vancampfort, D., Rosenbaum, S., Firth, J., Cosco, T., Veronese, N., Salum, G. A., & Schuch, F. B. (2017). An examination of the anxiolytic effects of exercise for people with anxiety and stress-related disorders: A meta-analysis. *Psychiatry Research*, *249*, 102–108. https://doi.org/10.1016/j.psychres.2016.12.020

Vijayan, V., Connolly, J. P., Condell, J., McKelvey, N., & Gardiner, P. (2021). Review of wearable devices and data collection considerations for connected health. *Sensors*, *21*(16), 5589.

Whoop. (2015, September 23). *WHOOP closes $12M, Announces Availability of its Performance Optimization System Exclusively Designed for Elite Athletes and Teams* [Press release]. https://www.whoop.com/us/en/press-center/whoop-closes-12m-announces-availability-of-its-performance-optimization-system-exclusively-designed-for-elite-athletes-and-teams/?srsltid=AfmBOopGjmnsZjENn_jl9-IALgSHwoI_1c15B6agXAZ-GcB5ksGwSNZs

Wisbey, B., Montgomery, P. G., Pyne, D. B., & Rattray, B. (2010). Quantifying movement demands of AFL football using GPS tracking. *Journal of Science and Medicine in Sport*, *13*(5), 531–536.

3

GAMIFICATION OF FITNESS AND TRAINING

Manning Zhang

Introduction

Although physical exercise is beneficial for both physical and mental health, maintaining a regular fitness regime is never easy (Strain et al., 2024). Research indicates that nearly 63% of novice gym-goers either discontinue or reduce their exercise workload within the first year of membership (Gjestvang et al., 2020). Encouraging fitness adherence, therefore, becomes a key task for coaches, fitness practitioners, and public health professionals. One proposed strategy to address this challenge is to incorporate game elements into exercise routines. Gamification, defined as the incorporation of game elements into a nongame context, is a strategy used to promote the initiation and sustained engagement in desired behaviors (Deterding et al., 2011; Kim & Werbach, 2016). Although game-like elements have long been incorporated into physical activities, advancements in technology, such as Bluetooth and smartphones, have significantly contributed to the rise of gamified fitness. From Nintendo fitness games in the 80s (Finco & Maass, 2014) to the latest VR-based Yoga apps (e.g., "YogaScape"), technology has continuously impacted how we engage with fitness. This chapter reviews the evolution of gamified fitness and synthesizes interdisciplinary research that examines its mechanisms, trends, and limitations.

Background

Gamification is a strategy of designing activities to make them more engaging and motivating by using game elements (Deterding et al., 2011). It has been utilized in various areas, such as education (Caponetto et al., 2014), healthcare (Brigham, 2015), business (Vardarlier, 2021), and food delivery services (van Doorn & Chen, 2021). Gamified fitness refers to the use of a gamification

DOI: 10.4324/9781003588405-4

strategy in physical activity service or product design (Lister et al., 2014). Based on how game elements are incorporated, gamified fitness products can range extensively from exergames (fitness-themed games such as "*Just Dance*") to fitness trackers (e.g., "Fitbit").

The history of gamified fitness traces back to the late 20th and early 21st centuries (Finco & Maass, 2014). In the 1980s, Nintendo, the well-known video game company, recognized that the public held a stigma that games encourage sedentary and unhealthy life habits (Finco & Maass, 2014). To alter this stereotype and, more importantly, to attract potential consumers, they started incorporating fitness elements into their game designs (Finco & Maass, 2014). One of their early products was "World Class Track Meet," an Olympic-themed video game released in 1986 (NintendoWiki, 2023). It came with a plastic mat called the "Power Pad," which Nintendo acquired and rebranded from the Japanese toy manufacturer Bandai (StrategyWiki, 2021). The classic version of the "Power Pad" has two sides, each featuring multiple red and blue buttons with sensors embedded. Users need to stomp on these buttons to control the game characters and complete fitness tasks like the long jump, triple jump, and 100-meter dash. Later dance exergames, like "Dance Dance Revolution," carried forward the sensor-controlled mat component (Ellis, 2019). These rhythm-based dancing games especially appealed to young women, a group historically underrepresented in the video game market. By providing a more accessible and social style of play, dance exergames drew in female consumers to a game market long dominated by traditionally masculine genres like action, sports, and combat (Finco & Maass, 2014).

In the 2000s, the exergame industry saw continued expansion. The release of the "Wii Fit" game with the "Wii" console in 2007, featuring various mini-exercise games like yoga and weight training, ushered in a new era for interactive fitness games (Millington, 2017, pp. 46–48). Meanwhile, the development of Bluetooth technology in 2004 and the launch of the first generation of smartphones paved the way for wearable tech and fitness apps (Williams, 2023). Early fitness-themed mobile phone apps like "Fit Phone" (Heffernan, 2022) and "Lose it!" (Anon, 2022) became available in mobile app stores, while wearable devices like "Fitbit" and "Jawbone UP" (Gay & Leijdekkers, 2015) gained popularity. The primary function of these early gamified fitness products was to track physiological data, such as weight, heart rate, steps, and food intake. The market for gamified fitness products flourished during the 2010s. The tech industry claimed that 2014 was "the year of the wearable" (Adapa et al., 2018; Spence, 2013). Trendsetters like Google, Samsung, and Sony strengthened the promotion of wearable devices, and smaller manufacturers entered the market (Gay & Leijdekkers, 2015). As of 2015, there were about 165,000 mobile health-related apps on Android and iOS platforms, with nearly 110,000 specifically targeting fitness and wellness (Gay & Leijdekkers, 2015; IMS Health, 2024).

By 2023, the revenue of the fitness apps market in the United States reached 1.51 billion dollars (Statista, 2024a), while the smart fitness tracker market reached 9.61 billion dollars (Statista, 2024b). Notably, the development and use of fitness apps grew significantly during and after the COVID-19 pandemic, driven in part by stay-at-home orders and increased health awareness globally (Liu et al., 2022; Tong et al., 2022).

Since the 1980s, gamified fitness products have become increasingly accessible and seamlessly integrated into people's everyday lives. Plenty of gamified fitness products are adapted to mobile phones and wearables and serve populations with diverse needs. From "Power Pad" to smartphone exercise apps, the gamification mechanisms buttressing gamified fitness products have remained similar over the years. In the next section, we will explore the key concepts in these mechanical processes.

Topical Exploration and Key Concepts

The concept of *gamification* was first coined in 2008 by Bret Terrill, who defined it as "taking game mechanics and applying them to other web properties to increase engagement" (Terrill, 2008). A later and well-embraced definition was proposed three years later by Deterding and his colleagues (2011), as "[the] use of game design elements in nongame contexts." Gamification refers to a process that intentionally adds a game element, such as "fun, play, transparency, design, and challenge" (Palmer et al., 2012, p. 54), to scenarios where games are not the default. The goal of gamification is to make human–product interaction more enjoyable (Zichermann & Cunningham, 2011) and enhance users' overall value-creation experiences (Huotari & Hamari, 2017). This framework is commonly used in marketing practices and product design for its ability to increase user engagement and boost the sales of the products (Jang et al., 2018). According to the projection of Markets and Markets (2020), the gamification market, across different domains, is set to grow rapidly from $9.1 billion in 2020 to $30.7 billion by 2025—more than tripling in just five years.

Gamification is not equal to game design. Robson et al. (2015) summarize gamification with the MDE model—mechanics, dynamics, and emotions—as an update to the MDA model (mechanics, dynamics, and aesthetics) used in traditional game design (Hunicke et al., 2004). Specifically, they argue that in the game design context, "aesthetics" refers to users evoking a wide range of emotions (such as submission, fantasy, sensation, etc.) when interacting with the machine (usually computers) (Hunicke et al., 2004). However, gamified products, compared to traditional games, usually target specific behavioral outcomes, which makes the open-ended dimension of aesthetics less relevant (Robson et al., 2015). As a result, the updated MDE model prioritizes emotions instead of aesthetics when conceptualizing gamification principles (Robson et al., 2015).

In this model, *mechanics* refer to the specific designer-made decisions to realize their goals. These decisions are reflected in the set up (set-up mechanics), the rule-setting (rule mechanics), and the reinforcement of the user behaviors (progression mechanics) (Robson et al., 2015). Numerous gamified fitness products compete in the market, but successful ones often share similar features that motivate users to stay engaged and continue using the service. Two major mechanics of gamification are *reward* and *challenge* (Tobon et al., 2020). People who use gamified fitness products usually have personal fitness goals, but it's often difficult to stick to a routine through self-discipline alone. The reward systems in these products are designed to keep users motivated by regularly reminding them of their goals and the benefits they can gain by completing tasks (Tobon et al., 2020). Helmefalk (2019) finds that gamified products and services of different disciplines commonly use "points" and "badges" to visualize rewards. These reward objects build a "loop" for the users to develop desired behaviors (Helmefalk, 2019). Regarding reward effectiveness, some researchers found that incommensurate game elements, like social media "likes" and "thumbs up," foster greater intrinsic motivation for fitness app users to stay engaged in the program compared to commensurate elements, such as points, that directly align with task performance (Feng et al., 2020).

Different from rewards, *challenges* motivate users by creating problems to be resolved. Csikszentmihalyi and LeFevre (1989, p. 816) posit that the most fulfilling experiences arise when individuals face significant challenges that correspond to their advanced abilities. This harmonious alignment not only brings joy but also creates a state called "flow" (Csikszentmihalyi & LeFevre, 1989). According to their theory, the "flow" drives personal growth by fostering skill development, boosting confidence, and enriching one's sense of complexity (Csikszentmihalyi & LeFevre, 1989). Goal-setting and competitions are typical forms of challenge mechanics in gamified fitness. For instance, wearable fitness technology incorporates goal-setting features by creating doable tasks that introduce gradual changes in users' physical activity behavioral patterns (Johnson et al., 2016), which are found to create strong motivations to partake in current and future physical activities (Schaffarczyk & Ilhan, 2019). Competition elements, like ranks and leaderboards, engage users by fulfilling their desire for social recognition. As for its effectiveness, Buchem et al. (2015) reported that senior users find the competition features in wearable technology helpful in maintaining physical fitness. However, Chen and Pu (2014) found that cooperation, or a mix of cooperation and competition, was significantly more effective than competition alone in boosting physical activity among college students using a mobile tracking app. These findings emphasize the varying impact of competitive features across demographics and contexts, calling for tailored gamified fitness designs.

Users' behavioral and mental responses to the product mechanics, according to the MDE model, are described as the *dynamics* and *emotional* aspects of the gamification design (Robson et al., 2015). While the perspectives of mechanics primarily reflect the designers' vision of the overall product structure, dynamics, and emotions pinpoint the envisioned users' overall experience with the product or service. Due to a host of interpersonal factors, such as education, sports preference, and personality, users experience fitness products in different ways. For example, in their evaluation of Strava, a fitness networking app, Whelan and Clohessy (2020) found that cyclists motivated by social recognition were more prone to developing obsessive training behaviors and experiencing burnout. Instead, users who valued the reciprocal benefits in the fitness app community tended to develop harmonious passion, which led to a decrease in their burnout rate.

Incorporating user-centered perspectives into gamification design improves the personalization and contextual relevance of products, yet there remains a gap in operationalizing dynamics and emotions—more often than not, one's reaction to the product or service can be quite private and personalized. To measure individuals' subjective experiences with gamification, Högberg et al. (2019) created the Gameful Experience Questionnaire (GAMEFULQUEST), designed to standardize the assessment of personalized interactions with products or services. In recent years, this questionnaire has been adapted and used in different empirical studies that gauge the users' experience of gamified fitness apps (Hassan et al., 2020; Rehman et al., 2024; Xi et al., 2024). Using this measurement, researchers have been making efforts to understand the relationship between the perceived "gamefulness" of the gamified fitness service and the exercise outcomes of the users.

Trends and Current Issues

In recent years, advanced technologies like artificial intelligence (AI) and extended reality technology have been integrated into gamified fitness, especially in the domain of exercise-themed serious games (Eun et al., 2023; Tolks et al., 2024). A serious game is a product that incorporates game elements to achieve goals that are not pure entertainment, and it is commonly used in healthcare and education (Wattanasoontorn, 2013). These games are typically designed as fitness tools tailored for specific groups undergoing pre- or post-medical procedures, such as physical therapy after surgery. AI technologies help machines process information and assist in decision-making, offering new possibilities for recovery. Serious games powered by AI for medical rehabilitation and motion recovery are a recent development, especially for patients recovering from strokes, limb loss, and similar conditions (Tolks et al., 2024). In most of these applications, AI does not analyze game data in real-time but evaluates users' fitness performance after game sessions. For example, "Farming" is an AI-integrated serious game designed to help elderly adults improve both their cognitive and physical abilities (Eun et al., 2023). After each game, an AI-based algorithm assesses the user's performance and adjusts the difficulty level to match their abilities to promote sustainable, long-term improvement in the user's recovery (Eun et al., 2023).

Beyond serious games, AI algorithms also contribute to the personalization of gamified fitness products, like recommender systems that forward fitness-related information, catering to the users' needs and encouraging their long-term engagement (Zhao et al., 2020). The personalization mechanism is realized via algorithms that learn and predict users' preferences for exercise, such as the location, timing, type of activity, number of people involved in the session (individual or group), the situation, and the underlying purpose (Venkatachalam & Ray, 2022). With this information, fitness apps can serve as personalized coaches who craft gamified, individual training plans for the users. Companies like 3DLOOK have integrated AI-driven body scanning technology into fitness tracking devices to monitor users' health and use the data in their mobile app, "FitXpress," to generate personalized training plans (3DLOOK, 2024). Workplace fitness has become a prominent field of personalized fitness gamification. For instance, "Vantage Fit" is a fitness product that supports gamified corporate wellness campaigns. Using gamified features like leaderboards, personalized health diaries, and trophies, "Vantage Fit" allows corporate managers to design fitness programs for employees, such as step challenges and virtual coaching (Singh, 2024). Beyond workplace fitness, personalized gamified fitness products also include smartphone applications tailored to support women's physical health during pregnancy and postpartum. However, high-quality pregnancy wellness applications can be expensive, and the existing free programs rarely engage with evidence-based workout suggestions that are tailored to their female audience (Tinius et al., 2021).

The integration of extended reality technology, particularly augmented reality (AR), is also noteworthy in fitness gamification. Research shows that AR smartphone games like "Pokémon Go," which require players to leave their homes and explore real-world locations, have been effective in increasing physical activity among young people (Marquet et al., 2018). Additionally, AR-based gamified exercise programs have also demonstrated their efficacy in improving balance and muscle control among older adults (Chen et al., 2020; Jeon & Kim, 2020). For example, a Taiwanese research team used AR technology to create a virtual Tai Chi master and simplified the movements for older adults (Chen et al., 2020). Their study found that, compared to traditional Tai Chi practice, these participants benefited at least as much—if not more—in terms of improved coordination and muscle strength (Chen et al., 2020). In addition, the AR-based Tai Chi program was designed to offer personalized training plans, catering to the diverse needs of learners. This approach helps address the challenge of simplifying and practicing Tai Chi independently (Chen et al., 2020).

A recent study by Sun and Yuan (2024) further demonstrates the effectiveness of incorporating gamified AR content into fitness programs. They found that users identified more with the hedonic rather than the utility value of the AR-based gamified fitness products and were thus motivated to stay engaged with the exercise (Sun & Yuan, 2024). In other words, the interactive features of these AR-based gamified fitness products make training more "fun," and they transform routinized exercise schedules into enjoyable activities (Sun & Yuan, 2024). In addition to AR, virtual reality (VR), a technology that immerses users into a simulated environment, can also greatly enhance the fitness experience by adjusting user-perceived surrounding areas. Some classical exergames, such as Ubisoft's *Just Dance*, have released VR versions, and research reveals that VR exergames can positively impact users' mental and physical health (Sarupuri et al., 2024; Yang et al., 2022). They enrich repetitive exercises by incorporating immersive visual and auditory elements within a controlled environment, with no new physical risks introduced (Yang et al., 2022).

While gamified fitness products have been praised for sparking a fitness craze and innovatively encouraging exercise habits, some researchers have raised ethical concerns about their design. Woodcock and Johnson's (2018) critique of gamified work routines argues that such programs either impose an authoritarian framework by mandating "play" and "fun" within the workplace or exploit values like merit and efficiency to encourage workers to self-instrumentalize. Building on this critique, gamified fitness in the workplace, and even beyond, can be viewed as a way to rationalize employees' free time through competition and progress, promoting efficiency and aligning with work-oriented productivity objectives. However, this critique cannot adequately address gamified fitness programs designed to enhance the quality of life for people who are not bound by work routines, such as retired citizens. For them, gamified fitness can be an active and effective use of time (Chen et al., 2020; Jeon & Kim, 2020).

Another concern with gamifying daily routines is the potential threat to personal data privacy. Arora and Razavian (2021) noted that the integration of gamification in health and fitness tracking apps could compromise user data. Similarly, Collins and Marassi (2021) highlighted that in corporate fitness programs, employers can leverage the "health data" collected by fitness trackers to monitor employees' health, behavior, and locations. They argue that these practices infringe upon employees' data privacy and autonomy (Collins & Marassi, 2021). Moreover, because game elements in gamified fitness programs are deliberately designed to foster behavior changes, designers often employ mechanisms that can subtly influence users' emotions and psychological states. This can lead to unintended consequences, such as generating psychological addiction to the game or increased stress (Arora & Razavian, 2021; Bai et al.,

2024). For instance, the use of competitive features like leaderboards, badges, and constant reminders may create a sense of pressure, motivating users to continuously engage with the game at the cost of their well-being. Over time, such stress-embedded activities can develop into a form of dependence, where users feel compelled to maintain their progress or status, leading to emotional and mental strain (Arora & Razavian, 2021; Bai et al., 2024).

Due to these potential hazards, some researchers maintain a skeptical view of the development of gamification mechanisms by large (Arora & Razavian, 2021; Collins & Marassi, 2021; Woodcock & Johnson, 2018). It is yet unclear whether gamification is an equitable tool that closes the health gap of physical activity or reinforces the gender, age, and class inequalities that already exist in society. For instance, as most gamified fitness programs are embedded in digital platforms, will this create barriers for populations who don't have access to or are unfamiliar with these technologies? Moreover, although gamified fitness is proven to enhance users' experience in exercise, what leads people to choose such services from the beginning? Are there potential barriers that restrict individuals who lack knowledge of physical exercise or cannot afford entry-level gym classes? To address these questions, it is important to understand the demographics of the gamified fitness users. In the next section, I will present a brief case study as a practice piece to explore the intriguing gender dynamics within the gamified fitness field.

Practice in Action

Women have historically been underrepresented in the fitness world and often seen as less physically active than men (Moreno-Llamas et al., 2022), yet recent data shows that women use gamified fitness apps over 20% more than men (Lyu et al., 2024). While this trend may offer hope that gender disparities in physical activity are narrowing, research from the COVID-19 pandemic reveals a more complex reality: women were more likely to be sedentary and overeat, even as they spent more time on fitness apps (Robinson et al., 2021; Tong et al., 2022). This suggests that higher fitness app usage alone doesn't necessarily mean women are leading more active lifestyles or truly benefiting from the apps. Understanding the exercise motivation behind women's consumption of gamified fitness is hence becoming important.

A survey of 839 fitness app users in the United States found that women's continued use of products is motivated by the development of a new habit, while men's engagement is based on the expectation that the apps will significantly improve their fitness performance (Yang & Koenigstorfer, 2021). Another study found that women fitness app users are better motivated by social support rather than the *reward* and *challenge* elements, such as the badges earned through practices, to perform more physical activities (Yin et al., 2022). These findings on the gendered consumption of gamified fitness suggest that women's exercise may not be fully driven by specific health or fitness goals but by the process of exercise, the connectivity, and the stability provided by conducting habitual, routinized activities (Yang & Koenigstorfer, 2021; Yin et al., 2022). The MDE gamification framework (Robson et al., 2015) emphasizes rewards and challenges as key motivators for behavior change. However, these design choices may be shaped by underlying gendered assumptions, as empirical evidence suggests that competitive factors might not be the most effective motivators for women to develop exercise behaviors (Yang and Koenigstorfer, 2021; Yin et al., 2022; Molanorouzi et al., 2015).

In my ethnographic study on exercise motivation, several middle-class women expressed a preference for women-only gyms over large, all-gendered fitness centers. One participant explained, "It's just nicer—I don't have to hear all the grunting or see dumbbells left everywhere…" They added, "…I'd even rather follow workouts on my fitness app than go to a gym where men are constantly flexing in the mirrors." For companies dedicated to developing women-focused gamified fitness apps, these messages may sound hopeful. As women-only gyms and boutique yoga studios often carry hefty price tags, women-focused fitness apps could offer a far more accessible and cost-effective alternative for women of different income circumstances.

However, gender inequity in physical exercise cannot be solved if not addressing the sexist social structure, especially revealed through the gendered distribution of leisure time: When women's hours dedicated to paid or family work increase, their level of physical activity decreases (Doan et al., 2022). Conversely, for men, their physical activity is not hindered by the added work time, and family time investment seems to be a protective factor for them to exercise (Doan et al., 2022). More research can be done in this thread to better understand how fitness apps, or other gamified fitness products, can be improved to serve women better.

Summary

Over the past four decades, gamified fitness has flourished, evolving alongside technological advancements. This chapter has examined its transformation from a niche interest within gaming communities to a powerful force shaping mainstream fitness habits. Moving forward, researchers and practitioners must continue exploring user engagement with gamified fitness through a sociological lens. This means considering how factors like time constraints, cultural influences, and access inequality shape people's fitness behaviors. While gamification makes exercise more engaging and accessible, true equity in fitness requires broader, interdisciplinary efforts to ensure all individuals have the resources to benefit from these innovations.

Key Resources

- National Public Radio: "How to gamify your exercise to make it more enjoyable" https://www.npr.org/2022/12/28/1145854110/how-to-gamify-your-exercise-to-make-it-more-enjoyable
- American Psychology Association: "Reflections on the Gamification of Fitness" https://blog.apaonline.org/2023/02/06/reflections-on-the-gamification-of-fitness/
- BBC Business Daily: "The gamification of fitness" https://www.bbc.co.uk/sounds/play/w3ct5zd6
- Medium.com: "Gamified Fitness Apps in 2024: The Trend You Cannot Miss" https://hitberrygames.medium.com/gamified-fitness-apps-in-2023-the-trend-you-cannot-miss-b1640232e67a

References

3DLOOK. (2024). FitXpress—AI-powered next-level fitness progress tracking. *3DLOOK*. https://3dlook.ai/fitxpress/

Adapa, A., Nah, F. F.-H., Hall, R. H., Siau, K., & Smith, S. N. (2018). Factors influencing the adoption of smart wearable devices. *International Journal of Human–Computer Interaction, 34*(5), 399–409. https://doi.org/10.1080/10447318.2017.1357902

Anon. (2022, June 28). Lose it! *Boston Magazine*. https://www.bostonmagazine.com/best-of-boston-archive/2022/lose-it/

Arora, C., & Razavian, M. (2021). Ethics of gamification in health and fitness-tracking. *International Journal of Environmental Research and Public Health, 18*(21), 11052. https://doi.org/10.3390/ijerph182111052

Bai, Z., Xu, C., Ailikamujiang, A., & Sun, Y. (2024). The dark side of gamification: An experimental study on digital fitness apps. *Kybernetes*. https://doi.org/10.1108/K-03-2024-0792

Brigham, T. J. (2015). An introduction to gamification: Adding game elements for engagement. *Medical Reference Services Quarterly, 34*(4), 471–480. https://doi.org/10.1080/02763869.2015.1082385

Buchem, I., Merceron, A., Kreutel, J., Haesner, M., & Steinert, A. (2015). Gamification designs in wearable enhanced learning for healthy aging. In *2015 International Conference on Interactive Mobile Communication Technologies and Learning (IMCL)* (pp. 9–15). https://doi.org/10.1109/IMCTL.2015.7359545

Caponetto, I., Earp, J., & Ott, M. (2014, October). Gamification and education: A literature review. In *European Conference on Games Based Learning* (Vol. 1, p. 50). Academic Conferences International Limited, Genova Italy.

Chen, P.-J., Penn, I.-W., Wei, S.-H., Chuang, L.-R., & Sung, W.-H. (2020). Augmented reality-assisted training with selected Tai-Chi movements improves balance control and increases lower limb muscle strength in older adults: A prospective randomized trial. *Journal of Exercise Science & Fitness, 18*(3), 142–147. https://doi.org/10.1016/j.jesf.2020.05.003

Chen, Y., & Pu, P. (2014). HealthyTogether: Exploring social incentives for mobile fitness applications. *Proceedings of the Second International Symposium of Chinese CHI*, 25–34. https://doi.org/10.1145/2592235.2592240

Collins, P., & Marassi, S. (2021). Is that lawful? Data privacy and fitness trackers in the workplace. *International Journal of Comparative Labour Law and Industrial Relations*, *37*(1). https://kluwerlawonline.com/journalarticle/International+Journal+of+Comparative+Labour+Law+and+Industrial+Relations/37.1/IJCL2021003

Csikszentmihalyi, M., & LeFevre, J. (1989). Optimal experience in work and leisure. *Journal of Personality and Social Psychology*, *56*(5), 815–822. https://doi.org/10.1037/0022-3514.56.5.815

Deterding, S., Dixon, D., Khaled, R., & Nacke, L. (2011). From game design elements to gamefulness. In *Proceedings of the 15th International Academic MindTrek Conference on Envisioning Future Media Environments* (pp. 9–15). Tampere, Finland, 28–30 September 2011.

Doan, T., Yu, P., LaBond, C., Gong, C., & Strazdins, L. (2022). Time for physical activity: Different, unequal, gendered. *Journal of Health & Social Behavior*, *63*(1), 37–54. https://doi.org/10.1177/00221465211028910

Ellis, R. (2019, September 25). The (Dance Dance) revolution. *Medium*. https://medium.com/@rachelellis_52506/the-dance-dance-revolution-dc1250229f72

Eun, S.-J., Kim, E. J., & Kim, J. (2023). Artificial intelligence-based personalized serious game for enhancing the physical and cognitive abilities of the elderly. *Future Generation Computer Systems*, *141*, 713–722. https://doi.org/10.1016/j.future.2022.12.017

Feng, W., Tu, R., & Hsieh, P. (2020). Can gamification increase consumers' engagement in fitness apps? The moderating role of commensurability of the game elements. *Journal of Retailing and Consumer Services*, *57*, 102229. https://doi.org/10.1016/j.jretconser.2020.102229

Finco, M., & Maass, R. (2014). The history of exergames: Promotion of exercise and active living through body interaction. In *2014 IEEE 3rd International Conference on Serious Games and Applications for Health (SeGAH)* (pp. 1–6). https://doi.org/10.1109/SeGAH.2014.7067100

Gay, V., & Leijdekkers, P. (2015). Bringing health and fitness data together for connected health care: Mobile apps as enablers of interoperability. *Journal of Medical Internet Research*, *17*(11). https://doi.org/10.2196/jmir.5094

Gjestvang, C., Abrahamsen, F., Stensrud, T., & Haakstad, L. A. (2020). Motives and barriers to initiation and sustained exercise adherence in a fitness club setting—A one-year follow-up study. *Scandinavian Journal of Medicine & Science in Sports*, *30*(9), 1796–1805. https://doi.org/10.1111/sms.13736

Hassan, L., Xi, N., Gurkan, B., Koivisto, J., & Hamari, J. (2020). Gameful self-regulation: A study on how gamified self-tracking features evoke gameful experiences. In *Proceedings of the 53rd Annual Hawaii International Conference on System Sciences, HICSS 2020* (pp. 1103–1112). https://doi.org/10.24251/HICSS.2020.138

Heffernan, C. (2022, March 16). Guest post: History of fitness apps every fitness freak should know - Physical culture study. https://physicalculturestudy.com/2022/03/16/guest-post-history-of-fitness-apps-every-fitness-freak-should-know/

Helmefalk, M. (2019). An interdisciplinary perspective on gamification: Mechanics, psychological mediators and outcomes: Mechanics, mental mediators and outcomes. *International Journal of Serious Games*, *6*(1). https://doi.org/10.17083/ijsg.v6i1.262

Högberg, J., Hamari, J., & Wästlund, E. (2019). Gameful experience questionnaire (GAMEFULQUEST): An instrument for measuring the perceived gamefulness of system use. *User Modeling and User-Adapted Interaction*, *29*(3), 619–660. https://doi.org/10.1007/s11257-019-09223-w

Hunicke, R., LeBlanc, M., & Zubek, R. (2004, July). MDA: A formal approach to game design and game research. In *Proceedings of the AAAI Workshop on Challenges in Game AI* (Vol. 4, No. 1, p. 1722), San Jose, California.

Huotari, K., & Hamari, J. (2017). A definition for gamification: Anchoring gamification in the service marketing literature. *Electronic Markets*, *27*(1), 21–31. https://doi.org/10.1007/s12525-015-0212-z

IMS Health. (2024). Retrieved October 23, 2024, from http://www.theimsinstitute.org [WebCite Cache]

Jang, S., Kitchen, P. J., & Kim, J. (2018). The effects of gamified customer benefits and characteristics on behavioral engagement and purchase: Evidence from mobile exercise application uses. *Journal of Business Research*, *92*, 250–259. https://doi.org/10.1016/j.jbusres.2018.07.056

Jeon, S., & Kim, J. (2020). Effects of augmented-reality-based exercise on muscle parameters, physical performance, and exercise self-efficacy for older adults. *International Journal of Environmental Research and Public Health*, *17*(9), 3260. https://doi.org/10.3390/ijerph17093260

Johnson, D., Deterding, S., Kuhn, K.-A., Staneva, A., Stoyanov, S., & Hides, L. (2016). Gamification for health and wellbeing: A systematic review of the literature. *Internet Interventions*, *6*, 89–106. https://doi.org/10.1016/j.invent.2016.10.002

Kim, T. W., & Werbach, K. (2016). More than just a game: Ethical issues in gamification. *Ethics and Information Technology*, *18*, 157–173.

Lister, C., West, J. H., Cannon, B., Sax, T., & Brodegard, D. (2014). Just a fad? Gamification in health and fitness apps. *JMIR Serious Games*, *2*(2), e3413. https://doi.org/10.2196/games.3413

Liu, R., Menhas, R., Dai, J., Saqib, Z. A., & Peng, X. (2022). Fitness apps, live streaming workout classes, and virtual reality fitness for physical activity during the COVID-19 lockdown: An empirical study. *Frontiers in Public Health*, *10*. https://doi.org/10.3389/fpubh.2022.852311

Lyu, L., Abidin, N. E. Z., & Zulnaidi, H. (2024). Exploring female fitness app users' motivations and perceptions: A qualitative study. *Medicine*, *103*(44), e40076. https://doi.org/10.1097/MD.0000000000040076

Markets and Markets. (2020). Gamification market growth drivers & opportunities. *Markets and Markets*. Retrieved January 6, 2025, from https://www.marketsandmarkets.com/Market-Reports/gamification-market-991.html

Marquet, O., Alberico, C., & Hipp, A. J. (2018). Pokémon GO and physical activity among college students. A study using ecological momentary assessment. *Computers in Human Behavior*, *81*, 215–222.

Millington, B. (2017). *Fitness, Technology and Society: Amusing Ourselves to Life* (1st ed.). Routledge.

Molanorouzi, K., Khoo, S., & Morris, T. (2015). Motives for adult participation in physical activity: Type of activity, age, and gender. *BMC Public Health*, *15*(1), 66. https://doi.org/10.1186/s12889-015-1429-7

Moreno-Llamas, A., García-Mayor, J., & De la Cruz-Sánchez, E. (2022). Gender inequality is associated with gender differences and women participation in physical activity. *Journal of Public Health*, *44*(4), e519–e526. https://doi.org/10.1093/pubmed/fdab354

NintendoWiki. (2023, April 29). World-class track meet. *StrategyWiki*. https://strategywiki.org/wiki/World_Class_Track_Meet

Palmer, D., Lunceford, S., & Patton, A. J. (2012). The engagement economy: How gamification is reshaping businesses. *Deloitte Review*, *11*, 52–69.

Rehman, U., Abbasi, A. Z., Ting, D. H., Hassan, M., & Khair, N. (2024). Exploring the impact of gamified experiences on user engagement in fitness apps: A GAMEFULQUEST perspective. *IEEE Transactions on Engineering Management*, *71*, 3613–3628. https://doi.org/10.1109/TEM.2023.3347231

Robinson, E., Boyland, E., Chisholm, A., Harrold, J., Maloney, N. G., Marty, L., Mead, B. R., Noonan, R., & Hardman, C. A. (2021). Obesity, eating behavior and physical activity during COVID-19 lockdown: A study of UK adults. *Appetite*, *156*, 104853. https://doi.org/10.1016/j.appet.2020.104853

Robson, K., Plangger, K., Kietzmann, J. H., McCarthy, I., & Pitt, L. (2015). Is it all a game? Understanding the principles of gamification. *Business Horizons*, *58*(4), 411–420. https://doi.org/10.1016/j.bushor.2015.03.006

Sarupuri, B., Kulpa, R., Aristidou, A., & Multon, F. (2024). Dancing in virtual reality as an inclusive platform for social and physical fitness activities: A survey. *The Visual Computer*, *40*(6), 4055–4070. https://doi.org/10.1007/s00371-023-03068-6

Schaffarczyk, L., & Ilhan, A. (2019). Healthier life and more fun? Users' motivations to apply activity tracking technology and the impact of gamification. In G. Meiselwitz (Ed.), *Social Computing and Social Media: Communication and Social Communities* (pp. 124–136). Springer International Publishing. https://doi.org/10.1007/978-3-030-21905-5_10

Singh, R. (2024, July 25). Gamification for employee fitness and corporate wellness. *Vantage Fit - A Global Employee Wellness Platform for Corporates*. https://www.vantagefit.io/blog/gamification-for-employee-fitness-and-corporate-wellness/

Spence, E. (2013). 2014 will be the year of wearable technology. *Fitday.com*. https://www.fitday.com/

Statista. (2024a). Fitness apps—Worldwide | Statista market forecast. *Statista*. https://www.statista.com/outlook/hmo/digital-health/digital-fitness-well-being/health-wellness-coaching/fitness-apps/worldwide

Statista. (2024b). Fitness trackers - United States. *Statista*. https://www.statista.com/outlook/hmo/digital-health/digital-fitness-well-being/fitness-trackers/united-states

Strain, T., Flaxman, S., Guthold, R., Semenova, E., Cowan, M., Riley, L. M., Bull, F. C., Stevens, G. A., Raheem, R. A., Agoudavi, K., Anderssen, S. A., Alkhatib, W., Aly, E. A. H., Anjana, R. M., Bauman, A., Bovet, P., Moniz, T. B., Bulotaitė, G., Caixeta, R., … Zoma, L. R. (2024). National, regional, and global trends in insufficient physical activity among adults from 2000 to 2022: A pooled analysis of 507 population-based surveys with 5·7 million participants. *The Lancet Global Health*, *12*(8), e1232–e1243. https://doi.org/10.1016/S2214-109X(24)00150-5

StrategyWiki. (2021, October 17). Power pad. *StrategyWiki*. https://strategywiki.org/wiki/Category:Power_Pad

Sun, Y., & Yuan, Z. (2024). A virtual gym in your pocket: The influence of augmented reality exercise app characteristics on user's continuance intention. *Virtual Reality*, *28*(1), 54. https://doi.org/10.1007/s10055-024-00959-4

Terrill, Bret. 2008. My coverage of lobby of the social gaming summit. Retrieved January 6, 2025, from http://blog.bretterrill.com/2008/06/my-coverage-of-lobby-of-social-gaming.html

Tinius, R. A., Polston, M., Bradshaw, H., Ashley, P., Greene, A., & Parker, A. N. (2021). An assessment of mobile applications designed to address physical activity during pregnancy and postpartum. *International Journal of Exercise Science*, *14*(7), 382–399. https://doi.org/10.70252/AQIG9215

Tobon, S., Ruiz-Alba, J. L., & García-Madariaga, J. (2020). Gamification and online consumer decisions: Is the game over? *Decision Support Systems*, *128*, 113167. https://doi.org/10.1016/j.dss.2019.113167

Tolks, D., Schmidt, J. J., & Kuhn, S. (2024). The role of AI in serious games and gamification for health: Scoping review. *JMIR Serious Games*, *12*. https://doi.org/10.2196/48258

Tong, H. L., Maher, C., Parker, K., Pham, T. D., Neves, A. L., Riordan, B., Chow, C. K., Laranjo, L., & Quiroz, J. C. (2022). The use of mobile apps and fitness trackers to promote healthy behaviors during COVID-19: A cross-sectional survey. *PLOS Digital Health*, *1*(8), e0000087. https://doi.org/10.1371/journal.pdig.0000087

van Doorn, N., & Chen, J. Y. (2021). Odds stacked against workers: Datafied gamification on Chinese and American food delivery platforms. *Socio-Economic Review*, *19*(4), 1345–1367. https://doi.org/10.1093/ser/mwab028

Vardarlier, P. (2021). Gamification in human resources management: A proposal for incorporating gamification in HRM. *Research Journal of Business and Management*, *8*(2), 129–139. https://doi.org/10.17261/Pressacademia.2021.1402

Venkatachalam, P., & Ray, S. (2022). How do context-aware artificial intelligence algorithms used in fitness recommender systems? A literature review and research agenda. *International Journal of Information Management Data Insights*, *2*(2), 100139. https://doi.org/10.1016/j.jjimei.2022.100139

Wattanasoontorn, V., Boada, I., García, R., & Sbert, M. (2013). Serious games for health. *Entertainment Computing*, *4*(4), 231–247. https://doi.org/10.1016/j.entcom.2013.09.002

Whelan, E., & Clohessy, T. 2020. How the social dimension of fitness apps can enhance and undermine wellbeing: A dual model of passion perspective. *Information Technology & People 34*(1), 68–92. https://doi.org/10.1108/ITP-04-2019-0156

Williams, J. (2023, April 10). The history of wearable tech. *BlumeHive*. https://www.blumehive.co.uk/post/the-history-of-wearable-tech

Woodcock, J., & Johnson, M. R. (2018). Gamification: What it is, and how to fight it. *The Sociological Review*, *66*(3), 542–558. https://doi.org/10.1177/0038026117728620

Xi, N., Buruk, O. 'Oz,' Chen, J., Jabari, S., & Hamari, J. (2024). Wearable gaming technology: A study on the relationships between wearable features and gameful experiences. *International Journal of Human-Computer Studies*, *181*, 103157. https://doi.org/10.1016/j.ijhcs.2023.103157

Yang, J., Menhas, R., Dai, J., Younas, T., Anwar, U., Iqbal, W., … Muddasar Saeed, M. (2022). Virtual Reality Fitness (VRF) for behavior management during the COVID-19 pandemic: a mediation analysis approach. *Psychology Research and Behavior Management*, 15, 171–182. DOI: 10.2147/PRBM.S350666.

Yang, Y., & Koenigstorfer, J. (2021). Determinants of fitness app usage and moderating impacts of education-, Motivation-, and gamification-related app features on physical activity intentions: Cross-sectional survey study. *Journal of Medical Internet Research*, *23*(7), e26063. https://doi.org/10.2196/26063

Yin, Q., Li, L., Yan, Z., & Guo, C. (2022). Understanding the effects of self-peer-platform incentives on users' physical activity in mobile fitness apps: The role of gender. *Information Technology & People*, *35*(3), 1054–1072. https://doi.org/10.1108/ITP-10-2020-0705

Zhao, Z., Arya, A., Orji, R., & Chan, G. (2020). Effects of a personalized fitness recommender system using gamification and continuous player modeling: System design and long-term validation study. *JMIR Serious Games*, *8*(4), e19968. https://doi.org/10.2196/19968

Zichermann, G., & Cunningham, C. (2011). *Gamification by Design: Implementing Game Mechanics in Web and Mobile Apps* (1st ed.). O'Reilly Media.

4

SPORTS ANALYTICS AND DATA-DRIVEN PERFORMANCE

Andrew Baerg

Introduction

Contemporary sport increasingly employs advanced statistical measurements to assess athletic performance. Whether it be "Next Gen Stats" like a quarterback's completion probability percentage in an National Football League (NFL) game, a hockey goalie's high-danger save percentage in an National Hockey League (NHL) playoff series, or a shooting guard's true shooting percentage across a season in the National Basketball Association (NBA), sports team front offices and sports fans now have access to more and more of these types of metrics. The use and public recognition of acronyms like wins above replacement (WAR), player efficiency rating (PER), and expected goals (xG) speak to the degree to which data-driven analytics has become part of conventional sports discourse. The sports industry's data-centric focus is often considered in the context of measuring and assessing athletic performance within the relatively new field of sports analytics. Szymanski (2020) defines sports analytics as "the use of performance data to understand the structure of sporting competition with the purpose of obtaining a competitive edge" (p. 57). Yet, even as much of the focus of sports analytics has been on the field for the past two decades (Awad, 2010; Curtis, 2015; Schwarz, 2005; Sheps, 2020), the sports industry is increasingly applying analytics to fan activity off the field. As a result, sports analytics and data-driven performance are no longer oriented solely around maximizing athlete performance (Ozkul & Duman, 2022). Sports organizations now seek to use analytics to maximize fan revenue from what has been deemed "sports venue analytics" as well (*Sports venue analytics: Leveraging data for fan engagement and operations*, 2024).

With advances in technologies (Mason, 2006; Peterson, 2021; Schaefer, 2023) that enable the tracking and calculation of athletic performance, analytics has

DOI: 10.4324/9781003588405-5

grown into a nearly ubiquitous element of twenty-first-century sport (Watanabe et al., 2021). Athlete measurement yields quantitative data points that feed into technologies supporting Big Data. Big Data represents "a capacity to search, aggregate, and cross-reference large data sets" (boyd & Crawford, 2012, p. 663). For sport, as in other areas of culture, Big Data's capacity allows for previously inaccessible correlations to be made and action then taken via data-based findings (Andrejevic, 2013). Sports analytics thereby combines new data collection technologies, new forms of data the technologies afford, and new ways to process these data in Big Data so that athletes and teams can be more competitive.

The chapter begins by briefly addressing the history of sports analytics and data-driven performance. It then discusses the concepts of analytics, Big Data, and the Quantified Self, concepts that inform further discussions of quantification and athletic performance, athletes and their relationship to data (especially in the context of privacy and data management), and the relationship between analytics and sports fans. The chapter concludes with a brief case study of the new home of basketball's Los Angeles Clippers, the Intuit Dome, as a space where analytics measure player activity on the court and fan activity off the court. To study sports analytics within its technological and cultural context is to better understand the way numbers and data shape our everyday experience. Following Beer (2016), examining sports analytics demonstrates how data has become a taken-for-granted aspect of how we relate to the world.

Background

While quantification and numbers are an essential aspect of modern sport (Guttmann, 1978), the roots of sports analytics and data-driven performance are comparatively recent. Although it is possible to look back further to the 1950s with the likes of baseball's Branch Rickey using more advanced forms of quantification to improve his Brooklyn Dodgers (Rickey, 1954) and Mottley (1954) using his military experience to emphasize a probability-based approach to sport, much of the sports analytics we see today originate in the 1970s with the advent of advanced baseball statistics like On-Base Percentage, Runs Created, and WAR (Puerzer, 2002; Schwarz, 2004). The desire for more advanced forms of statistical measurement would spread to other sports in the late twentieth century and become widespread into the twenty-first century (Baerg, 2025).

Why complex forms of quantification should arise is an important question. Beyond technological advances in computing power and the calculation speeds they allowed, we could also turn to the late twentieth-century neoliberal context in which these developments occurred. Following the decline of the welfare state and its cradle-to-grave intervention in the lives of its citizens, neoliberal states emphasize governing through freedom such that neoliberal citizens must be responsible for themselves and their futures (Andrews & Silk, 2018; Rose & Miller, 2010). Although numbers in government have a longer historical arc (Crosby, 1997; Hacking, 1990; Porter, 2011), numbers serve as a particularly important technique by which the neoliberal citizen exercises the responsibility of self-government. As neoliberal citizens measure, calculate, and learn to think about the world through a quantitative lens, citizens manage their presents and futures. Neoliberal citizens use numbers to manage existing and potential risk in a time when they cannot rely on the government to intervene (Dean, 1999; Miller & Rose, 2008; Rose, 1999; Shore & Wright, 2018).

A preoccupation with a quantitatively defined future contributes to the increased use of numbers in sports and the explosive growth of sports analytics. The ubiquity of sophisticated statistics in contemporary sport leads Beer (2016) to assert sport as an expression of "everyday neoliberalism" (p. 2). As such, we see this everyday neoliberalism playing itself out in sports analytics that regularly shapes front office decisions and player performance. This shaping occurs in sports as diverse as football (Pappalardo et al., 2019), curling (Chan & Golen, 2022), and cricket (Vishwarupe et al., 2022), among a vast range of others. In a neoliberal context, those employing analytics in sport measure the present in hopes of influencing the future. This future orientation increasingly shapes the sports fans' experience with sport as well. The formal adoption of sports analytics in professional sport occurs alongside the related informal, numbers-oriented fan activities that are fantasy sports (Baerg, 2016) and sports gambling (Franks et al., 2016). As with sports organizations' use of analytics, fans who draft a fantasy team or gamble on a game employ a quantitative epistemology in the present to predict future sporting outcomes.

Topical Exploration and Key Concepts

Sport provides a public-facing, visible expression of the nature of analytics and how they function (Beer, 2016). Perhaps the most visible example of this publicity arrived with Michael Lewis' 2003 book, *Moneyball*, and its film version starring Brad Pitt in 2011. The book and the film chronicle how baseball's Oakland Athletics used analytics to create a winning team on a shoestring budget. Other less-visible developments in sports analytics involve Massachusetts Institute of Technology's (MIT's) Sloan Sports Analytics Conference, university programs devoted to sports analytics, and the advent of the *Journal of Quantitative Analysis in Sports*. The now-annual Sloan Conference began in 2006 as a hub of innovation for the field. The conference features a variety of speakers making sports analytics presentations for an audience of "industry professionals (executives and leading researchers), students, and fans" (SSAC, 2024, para. 1). Academic institutions have developed sports analytics degrees with at least 25 American schools now offering sports analytics as an undergraduate major.

Academic interest in sports analytics can also be witnessed in the ongoing publication of the *Journal of Quantitative Analysis in Sports*. Begun in 2005, the journal's most recent volumes have published articles on topics like positional plus-minus in the NBA, batting orders in baseball, and sports betting strategies. Like the Sloan Conference, the journal continues to support the growth of new ideas in a burgeoning discipline.

Sports analytics is ubiquitous in professional sport but is now a part of amateur sport as well. At the amateur level, an app like Garmin Connect tracks a variety of fitness-related variables. To be a casual cyclist using a Garmin Edge cycling computer in conjunction with Connect is to be able to track everything from average heart rate to maximum cadence to the power generated by each leg. Garmin then combines these data points into ride performance and fitness metrics. Select data allow riders to compare themselves to other cyclists and others in their age bracket. Each ride's metrics correlate in Garmin's analytics algorithms to yield new data fields like training session intensity and fitness age scores. Analytics are even finding their way into youth sports. Little League Baseball endorses an official app, GameChanger, that has the capacity to track advanced metrics for children as young as four years old (Sanderson & Baerg, 2020).

Big Data

Sports analytics becomes ubiquitous with the advent of Big Data. Concisely defining Big Data is challenging, but boyd and Crawford (2012) provide an oft-cited understanding of Big Data as the nexus of technology, analysis, and mythology. Each of these areas shapes the way Big Data comes to be employed in the context of sport. First, Big Data requires sophisticated computing technology to collect, aggregate, store, and link large data sets. Computing technologies are also needed to deploy the algorithms that sort and parse this data. With advances in computing technology, Big Data also allows for the processing of massive amounts of information in real time. Second, Big Data enables new ways to analyze data and new conclusions to be drawn from this analysis. With its ability to identify correlations that would otherwise remain invisible, Big Data allows for new arguments to be made about the world and how we ought to act within it. Third, Big Data brings with it a mythology that assumes that the knowledge it produces is truthful and free of subjectivity. Failing to draw on this new access to Big Data's objective truth is to be deemed less intelligent or partial (boyd & Crawford, 2012; Hutchins, 2016).

Modern sports have consistently been interested in collecting quantitative data. For example, baseball has been preoccupied with using numbers to assess players and player performance almost since its inception (Schwarz, 2004). However, the twenty-first-century sports' employment of Big Data technology digitizes, accelerates, and hypercharges this preoccupation. New visual technologies allow for real-time tracking of athletic movement in fine-grained detail. As an example, the NBA's "Dunk Score" metric charts location data "for 29 different points on all 10 players' bodies 60 times per second" (*What is the 'Dunk Score?'*, 2024, p. para.8) to assess a dunk's value based on how far the dunker was from the basket, how high he jumped, how quickly the ball went through the hoop, and whether or not a defender attempted to block it. This data is then added to a model that calculates which player had the best dunk over a night of games or over the season. Incredibly, the tracking system collects, aggregates, and stores 17,400 pieces of data every second and more than one million data points each minute.

Modern sport's integration of Big Data has also fostered new arguments about player performance in sport. Under the influence of Big Data, older ideas about effective strategy and team composition have faded away. For example, decades of aggregated data have yielded a shift in baseball strategy as the home run and the strikeout have become the privileged offensive and defensive outcomes, respectively (Smith, 2024). Big Data has fostered a similar seismic shift in basketball strategy. What was once an inside strategic focus designed to get good shots at the rim has now shifted to tactics emphasizing three-point shots well away from the basket (Dawson, 2023). Big Data has also influenced American football's wage structure as teams have moved to a more efficient passing game that deemphasizes running backs. These running backs subsequently have less value than they once did and are subsequently paid much less relative to those who played the same position two and three decades earlier (Maaddi, 2023).

The mythology of Big Data expresses itself in sport by linking the use of sports analytics to an alleged objective intelligence that yields optimal performances, increased competitiveness, and winning. In its initial diffusion, sports analytics butted up against this mythology. Older scouts countered the mythology with forms of assessment epistemology emphasizing intangible, immeasurable qualities like fear or heart (Schwarz, 2005). Former athletes, like basketball's Charles Barkley, saw advanced statistical measures as the domain of "nerds" who had never played professional sports (Curtis, 2015). As such, analytics' objective emphasis and its attendant Big Data mythology were countered by those suggesting that the approach was hardly intelligent at all. Against this older "take it or leave it" approach to analytics, sports organizations lacking an analytics focus now face scrutiny from fans who believe these teams are not acting intelligently. Analytics now sit comfortably within a range of assessments of player performance.

Quantified Self

Sports analytics are not only deployed from above by sports institutions and organizations but have the capacity to shape individual behavior below from within the self. In his work on the body, Foucault (1984) argues that statistics render the body as "more or less utilizable, more or less amenable to profitable investment" and privilege "those with greater or less prospects of survival, death, and illness, and with more or less capacity for being carefully trained" (p. 279). While statistics would measure the body's utility as part of a larger welfare state, in neoliberalism, the self is encouraged to internalize these quantitatively inflected ideas to measure personal utility. Subsequently, this neoliberal person becomes "rendered calculable to others and to him- or herself in terms of numbers" (Rose, 1999, p. 213). In this context, a person's value is based on their ability to be productive, and this productivity is measured via calculation. Both individual value and how this value is assessed become important for understanding how the neoliberal self functions and performs. If quantifying the self becomes a means of measuring one's value, then considering the data the self produces and applying it to everyday activity becomes a logical next step. The Quantified Self collects quantitative data in the tracking of daily routines. The Quantified Self then uses collected "numbers as a means of monitoring elements of everyday life and embodiment" (Lupton, 2016, p. 3).

Although it is difficult to discern the specifics of how athletes have embraced a Quantified Self, it is clear that athletes have embraced new technologies to maximize their physical capital, that is, the value they can generate from their bodies. Basketball superstar LeBron James reportedly invests $1.5 million annually in order to "biohack" his body. James' carefully crafted biohacking regimen involves cryotherapy, red light therapy, sessions in a hyperbaric chamber, a sleep environment that maintains a prescribed temperature range, and a diet that counts carbohydrates and calories (Swartz, 2024). To routinely measure oxygen levels in the blood, muscle recovery times, hours and minutes of sleep, and calorie intake on a routine basis embodies the approach to everyday living adopted by the technologically invested Quantified Self.

Trends and Current Issues

Quantification and Athletic Performance

Most obviously, sports analytics and data quantify the athlete. As stated above, numbers have been vital to the development of modern sport and even important to pre-modern sport (Carter & Kruger, 1990). Numbers have been used for basic functions like keeping score and tracking athletic performance over time. Tulle (2016) explains how numbers have also been used to measure a given athlete's physical capital. In her research, she chronicles how sports media use numbers like age and experience to chart how an athlete might be aging and declining. Within sports organizations, sports analytics and an increasingly data-driven sports culture take these easily accessible numbers and add exponential sophistication and complexity to understanding an athlete's existing and potential physical capital. Organizations employ analytics for injury prevention to both safeguard this physical capital and optimize athletic productivity.

This quantification of the athlete raises several questions. One question that might be raised concerns the place of the aesthetic in athletic performance. Can there continue to be space for the aesthetic beauty of appreciating athletes regardless of what the data tell us? Sports analytics foregrounds measurable performance to such a degree that other aspects of sport could potentially be marginalized, if not eliminated entirely. The question of aesthetics may also be related to biomechanics as well. Athletes who move in unorthodox, inefficient ways may be extremely successful but pushed toward motion that has been deemed biomechanically efficient. Yet, these athletes' seemingly inefficient biomechanics may be precisely that which makes them successful. At the same time, this biomechanical unorthodox movement might end up being aesthetically pleasing for sports fans who have become inured by analytically informed, efficient biomechanics. As an example, basketball's "Euro step", introduced in the mid-1980s by Lithuanian Sarunas Marciulionis and popularized in the 2000s by Manu Ginobili, Giannis Antetokounmpo, and James Harden, is biomechanically inefficient in being both a slower and less direct way to get to the rim. Yet, the Euro step has been embraced by fans and players alike as an exciting, effective technique (Brenner, 2018).

Quantifying the athlete through sports analytics may also raise questions about athletes' humanity. To reduce an athlete to what that person measurably produces suggests that an athlete's dignity and value reside in the data they generate. Reducing athletes to data touches those in sports organizations who must make personnel decisions. But the question of dehumanization also increasingly touches sports fans involved in the likes of fantasy sports and sports gambling. Although existing research in this area is limited, Larkin (2022) has argued that dehumanization is more likely to occur in fans who have invested money in fantasy sports or in sports gambling. If money is at stake, these fans are more inclined to perceive athletes as objects in their service rather than humans worthy of agency and respect.

Using Athletes' Performance Data

A second issue related to sports analytics and athletic performance arises with legal and regulatory questions about data ownership, privacy, and curation. Myriad questions arise about what data can be collected, how long it can be held, to what purposes it might be put, and to whom it might be made available.

Although much of the discussion has revolved around sports analytics in the context of on-field productivity, on-field performance is being correlated to off-field biology and athlete activity. For roughly a decade, these questions have become increasingly focused as much on these off-the-field concerns as they are about on-field performance. In 2014, then Denver Nuggets general manager, Pete D'Alessandro, established his dream scenario for collecting off-the-court athlete data declaring that, "The holy grail is sequencing and understanding the genome. And how that relates to pro athletes on an injury basis and who's naturally good at certain sports" (Torre & Haberstroh, 2014, para. 41). In the same article, D'Alessandro shared his hope for a DNA database he could tap into before the NBA's annual draft to best assess which players would be most productive and injury-free ahead of his team's selection. Biometric information may not necessarily count as analytics in the purest sense; however, the prospect of uniting more and more biometric data with analytics intensifies the gravity of these data-related questions.

How sports analytics data has been handled serves as another issue in sports. Athletes have responded in different ways to the issue of data and sports analytics. Invariably, issues of data ownership and use are fraught with legal complexity about what data count as intellectual property and can be deemed proprietary and confidential, how owned data can be monetized, to what ends, and assigned what value. In North America, professional athletes have typically relied on unions to negotiate these concerns. In Europe, 1,400 member-athletes of the global athlete union, Federation Internationale des Associations de Footballeurs Professionels (FIFPRO), joined a 2020 class action suit against gambling and gaming companies that had used their performance data without compensating them. The athletes argued that their performance data should be equated to personal data and thereby subject to the same regulations approved by the European Union's 2016 General Data Protection Regulation (GDPR). GDPR provided considerably more extensive protections for citizens with respect to data collection and storage and, pertinent to this case, required businesses to compensate those from whom data would be collected for commercial purposes. As of this writing, FIFPRO has committed to creating "a data management system that will give individual players the ability to store and control their own performance data throughout their careers" (Slater, 2023, p. para.1), even as it remains unclear about what shape this system will take or how it will be implemented.

Regardless of whether athletes' data is freely accessible and/or used in proprietary databases, how the data should be treated and whether athletes should be compensated for the data they produce remains an ongoing issue.

Surveillance of Fans and Sports Venue Analytics

As stated in previous sections, sports analytics is no longer restricted to assessing athlete performance. Analytics has now spread to the assessment of what has been deemed "sports venue analytics" (*Sports venue analytics: Leveraging data for fan engagement and operations*, 2024) as well. As long as tickets have been sold for people to watch professional sport, sporting institutions and organizations have been interested in maximizing the revenue generated from live sport. These organizations have been able to gather more and more information on their customers over the past century, but recent technological innovations now mean that they can gather increasingly complex data on how fans engage with their respective products. As websites track fan activity through cookies and phone apps record every swipe of a fan's finger, leagues and teams can gauge how fans are performing as much as they gauge how athletes are performing. Surveillance of sports fans is a considerably complex research area (Whelan, 2014) but recent developments bring fan surveillance into contact with sports analytics and Big Data. Sports venue analytics takes the same approach to tracking athletic performance on the field and applies it to fan performance off of it.

Professional sports facilities are increasingly shifting toward facial recognition as a proxy for tickets. Facial recognition has been part of baseball stadium attendance since 2018 (Holms, 2018). In 2024, the NFL began offering fans the chance to use facial recognition as a ticket replacement. Brandon Covert, vice-president of information technology for the Cleveland Browns, has stated that he would like to see a Browns fan's face unlock entry into the stadium, enabling food and merchandise purchase (Stanley, 2024). Theoretically, this single access key makes the fan's life more convenient. No longer does the fan need to search for a physical ticket or pull a credit card out of a wallet for a hot dog or team jersey. Frictionless entry to a stadium and frictionless purchases inside the stadium conceivably make the fan's experience more efficient. Much like the athletes they are watching perform, so too is the fan's performance under increased supervision, data collection, and aggregation.

Efficient fan surveillance yields some issues to consider. Surveillance of fans can yield a Big Data digital divide (Andrejevic, 2014) whereby those under surveillance are unaware of which data collection tools are being used, what data is being collected, or how this collected data is being used (Kennedy & Moss, 2015). As with the player data mentioned above, this fan-generated data raises privacy concerns. Stanley (2024) and Tenbarge (2024) argue that facial recognition systems can also tend toward inaccuracy for non-whites. They also caution that facial recognition data can be co-opted by public law enforcement and private security to blacklist and ban fans who might be deemed problematic. In addition, they also assert that facial recognition data will likely be combined with other databases in increasingly centralized ways.

Undergirding each of these trends and issues sits the looming influence of Artificial Intelligence (AI). AI will undoubtedly shape how athletic performance is assessed by front offices, who might use it to quickly evaluate player performance. AI will be deployed by players and players' unions to better discern how they should be compensated by leagues and team owners. AI will be used to quickly aggregate information about fans, whether it be what they buy from a team's online storefront or how they might move within a stadium.

Practice in Action

Sports analytics are often most directly linked to athletes. Yet, as explained above, sports fans are increasingly the target for analytics in their consumption of sport. One of the most vivid examples of this targeting comes from the newly opened Intuit Dome, the home of basketball's Los Angeles Clippers. That the facility is sponsored by Intuit, a multinational corporation specializing in the quantitatively focused areas of tax preparation, accounting, and credit monitoring seems to be hardly a coincidence. The Intuit Dome's surveillance of athletes and fans throughout their experience with the facility illustrates some of the ideas outlined above.

The Intuit Dome's collection of player data for analytics comes from 14 Hawk-Eye cameras purchased from Sportsradar and stationed at locations standardized for every NBA arena. Hawk-Eye cameras track 29 points on players' bodies alongside the ball. The cameras collect data at a rate of 60 frames per second. The data is then fed into Sportsradar's Synergy Sports system, a system that uses algorithms to measure how conventional basketball plays like drives and picks are performed and defended. However, the Hawk-Eye system extends this measurement by linking this conventional data to new data on hand, limb, and foot placement. Hawk-Eye's body movement tracking allows for novel forms of training and practice as the data can be adapted for visualization exercises. Coaches can now show players where specific parts of their bodies were on a given play and generate a visual simulation of how that movement might be better optimized (Schaefer, 2023). Clippers' management, data scientists, and coaches can also take this information and generate their own proprietary models for assessing player performance, predicting a player's future value, and preventing injuries. The players may choose to use Hawk-Eye data as part of a Quantified Self for their own competitive and injury mitigation purposes.

Rudimentary analytics data collected from these cameras is passed along to attending fans. The Intuit Dome's massive scoreboard, nicknamed "The Oculus", features conventional information about the game being played, but also includes advanced statistics related to shot charts and game flow. Attentive fans can quickly get a more sophisticated understanding of what might be occurring in the game beyond the traditional performance data arenas provide that are related to points, assists, rebounds, steals, and blocks. Fans can employ this data instrumentally to perceive players as data producers as much as individuals with agency and emotion.

The Intuit Dome's data generation is not restricted to Clippers players and their opponents. The Intuit Dome also leverages AI and biometric technology to surveil fans in just about every area of the facility. To enter the arena, fans must download apps that provide them with access. One, Identity Pass, requires fans to enter profile information on Identity Pass in their phones, add Identity Pass to their Google or Apple wallets, and then tap stanchions near arena entrances using the app to get in. A second app, GameFace ID, requires sports fans to take a selfie and then upload the image to GameFace ID. Fans can subsequently walk directly into the arena without even needing to take their phones out of their pockets. To enter areas where they can purchase concessions or merchandise, sports fans are also required to use one of these apps. The apps perform surveillance on fans as they gather food or team merchandise. After picking up any desired items, fans are subsequently charged as soon as they leave the area and a receipt is submitted through the app. As such, cash is meaningless at the arena (De Leon, 2024). Identity Pass or GameFace ID is also needed to enter suites or club areas within the arena as well.

The Intuit Dome's surveillance of the sports fan ties to several of the issues alluded to above. Most prominently, the Intuit Dome collects myriad data on sports fans as they enter, circulate, and make purchases in the facility. In uniting venue access, concession purchases, and merchandise acquisition via facial recognition, the Clippers have achieved what the aforementioned Cleveland Browns would like to achieve. Fan activity is also tracked while fans engage with video game-like console controllers to respond to scoreboard prompts. Some areas of the arena even measure a given fan's decibel level as a measure of that person's enthusiasm.

The Intuit Dome provides an excellent example of how sports analytics shape the way sport operates, how fans experience sports, and how the future of sports may unfold for athletes and fans subject to an analytics approach to their activity.

Summary

This chapter has provided a brief history of sports analytics by situating its development in the historical context of neoliberalism and its emphasis on employing numbers as a tool of government. Sports analytics provides one of the most visible expressions of neoliberalism in its tracking, measurement, and assessment of athletes at the professional and amateur levels. Sports analytics also illustrates a mythology attached to Big Data in deploying ever-powerful computing technology to create new correlations between different data points, correlations that lead those using Big Data to suggest its use is intelligent and objective.

The chapter has also addressed a few of the key issues and trends related to sports analytics. First, sports analytics may marginalize the aesthetic aspect of sport performance while potentially dehumanizing the athletes it measures. Second, sports analytics raises ethical questions about data ownership, data privacy, and what athletes should and should not be able to do with the data they produce. Third, questions about sports analytics on the court spill over into sports venue analytics off the court as fan behavior begins to come under the same kind of surveillance as the athletes the fans come to watch. The chapter uses the newly built arena of the Los Angeles Clippers as a case study and a place where older and newer sports analytics practices come together.

Key Resources

- MIT Sloan Sports Analytics Conference: https://www.sloansportsconference.com/
- *Journal of Quantitative Analysis in Sports*: https://www.degruyter.com/journal/key/jqas/html
- Sportradar: https://sportradar.com/teams-leagues-federations/synergy-coaching-scouting/
- Garmin Connect: https://connect.garmin.com/
- Intuit Dome: https://intuitdome.com/

References

Andrejevic, M. (2013). *Infoglut: How Too Much Information Is Changing the way We Think and Know*. Routledge.

Andrejevic, M. (2014). The big data divide. *International Journal of Communication*, *8*, 1673–1689.

Andrews, D. L., & Silk, M. L. (2018). Sport and neoliberalism: An affective-ideological articulation. *The Journal of Popular Culture*, *51*(2), 511–533. https://doi.org/10.1111/jpcu.12660

Awad, T. (2010). Introduction to GVT and VUKOTA. In A. Rothstein & T. Seppa (Eds.), *Hockey Prospectus 2010–11* (pp. vii–x). Prospectus Entertainment Ventures LLC.

Baerg, A. (2016). Draft day: Risk, responsibility and fantasy football. In N. D. Bowman, J. S. W. Spinda, & J. Sanderson (Eds.), *Fantasy Sports and the Changing Sports Media Industry: Media, Players, and Society* (pp. 99–120). Lexington Books.

Baerg, A. (2025). Sport, analytics, and the number as a communication medium. In P. Pedersen (Ed.), *Routledge Handbook of Sport Communication* (2nd ed., pp. 373–381). Routledge.

Beer, D. (2016). Productive measures: Culture and measurement in the context of everyday neoliberalism. *Big Data & Society*, *2*(1). https://doi.org/10.1177/2053951715578951

boyd, d., & Crawford, K. (2012). Critical questions for big data. *Information, Communication & Society*, *15*(5), 662–679. https://doi.org/10.1080/1369118X.2012.678878

Brenner, J. (2018). Why the Eurostep is the NBA's most controversial move--and its most lethal. Retrieved May 29, 2025, from https://www.espn.com/nba/story/_/id/25206602/most-dangerous-two-step-game

Carter, J. M., & Kruger, A. (Eds.) (1990). *Ritual and Record: Sports Records and Quantification in Pre-Modern Societies*. Greenwood Press.

Chan, K., & Golen, J. (2022, February 18). Rock and ice meet big data: How Olympic curlers use stats to win. *The Christian Science Monitor*. https://www.csmonitor.com/World/2022/0218/Rock-and-ice-meet-big-data-How-Olympic-curlers-use-stats-to-win

Crosby, A. W. (1997). *The Measure of Reality: Quantification and Western Society, 1250–1600*. Cambridge University Press.
Curtis, B. (2015). Moneyball II: Charles Barkley, the sports media, and the second statistical war. *Grantland.com*. Retrieved August 5, 2016, from http://grantland.com/the-triangle/moneyball-advanced-statistics-charles-barkley-sports-media-daryl-morey-al-leiter-rob-neyer-nba-mlb-nfl-nhl/
Dawson, M. (2023). The iron cage of efficiency: Analytics, basketball and the logic of modernity. *Sport in Society*, *26*(11), 1785–1801. https://doi.org/10.1080/17430437.2023.2208040
De Leon, A. (2024, October 24). What's it like to attend a game at the Clippers' $2-billion arena? *Los Angeles Times*. https://www.latimes.com/sports/clippers/story/2024-10-23/clippers-intuit-dome-new-arena-amenities
Dean, M. (1999). *Governmentality: Power and Rule in Modern Society*. Sage Publications.
Foucault, M. (1984). The politics of health in the eighteenth century. In P. Rabinow (Ed.), *The Foucault Reader* (pp. 273–289). Pantheon.
Franks, A., D'Amour, A., Cervone, D., & Bornn, L. (2016). Meta-analytics: Tools for understanding the statistical properties of sports metrics. *Journal of Quantitative Analysis in Sports*, *12*(4), 151–165. https://doi.org/10.1515/jqas-2016-0098
Guttmann, A. (1978). *From Ritual to Record: The Nature of Modern Sports*. Columbia University Press.
Hacking, I. (1990). *The Taming of Chance*. Cambridge University Press.
Holms, E. (2018). MLB partners with Clear for biometric ticket roll-out. *Sportspro.com*. Retrieved May 29, 2025, from https://www.sportspro.com/news/mlb-clear-biometric-ticketing-facial-recognition-tech/
Hutchins, B. (2016). Tales of the digital sublime: Tracing the relationship between big data and professional sport. *Convergence*, *22*(5), 494–509. https://doi.org/10.1177/1354856515587163
Kennedy, H., & Moss, G. (2015). Known or knowing publics? Social media data mining and the question of public agency. *Big Data & Society*, *2*(2). https://doi.org/10.1177/2053951715611145
Larkin, B. (2022). Toward a better understanding of fantasy football and dehumanization of athletes: Is fantasy sport gambling to blame? *Sport Marketing Quarterly*, *31*(4), 292–304. https://doi.org/10.32731/SMQ.314.1222.04
Lupton, D. (2016). *The Quantified Self: A Sociology of Self-Tracking*. Polity Press.
Maaddi, R. (2023). NFL teams are refusing to pay running backs. *That reflects the pass-first direction of the league*. Retrieved May 30, 2025, from https://apnews.com/article/nfl-running-backs-value-saquon-barkley-josh-jacobs-c5d77e87911c1ab993d9ae312a9040ad
Mason, D. S. (2006). Moneyball as a supervening necessity for the adoption of player tracking technology in professional hockey. *International Journal of Sports Marketing & Sponsorship*, *8*(1), 47–61.
Miller, P., & Rose, N. (2008). *Governing the Present: Administering Economic, Social and Personal Life*. Polity Press.
Mottley, C. M. (1954). Letter to the editor--The application of operations research methods to athletic games. *Journal of the Operations Research Society of America*, *2*(3), 335–338. https://doi.org/10.1287/opre.2.3.335
Ozkul, A., & Duman, G. (2022). Analytics on service quality in spectator sports. *Journal of Applied Sport Management*, *14*(4), 48–53. https://doi.org/10.7290/jasm14h59c
Pappalardo, L., Cintia, P., Rossi, A., Massucco, E., Ferragina, P., Pedreschi, D., & Giannotti, F. (2019). A public data set of spatio-temporal match events in soccer competitions. *Scientific Data*, *2019*(6), 1–15. https://doi.org/10.1038/s41597-019-0247-7

Peterson, B. (2021, April 28). Football's tech-driven future of performance [Interview]. *SportTechie.* https://app.live?storm.co/sporttechie/footballs-tech-driven-future-of-performance/live?s=a504cbe6-cd28-4309-92b6-83b4b6f3fcdd#/chat

Porter, T. M. (2011). Statistics and the career of public reason: Engagement and detachment in a quantified world. In T. Crook & G. O'Hara (Eds.), *Statistics and the public sphere: Numbers and the people in modern Britain, c. 1800–2000* (pp. 32–47). Routledge.

Puerzer, R. J. (2002). From scientific baseball to sabermetrics. *NINE: A Journal of Baseball History*, *11*(1), 34–48.

Rickey, B. (1954, August 2). Goodbye to some old baseball ideas: 'The Brain' of the game unveils formula that statistically disproves cherished myths and demonstrates what really wins. *Life*, 78–89.

Rose, N. (1999). *Powers of Freedom: Reframing Political Thought*. Cambridge University Press.

Rose, N., & Miller, P. (2010). Political power beyond the state: Problematics of government. *The British Journal of Sociology*, *61*, 271–303.

Sanderson, J., & Baerg, A. (2020). Youth baseball and data analytics: Quantifying risk management and producing neoliberal responsible citizenship through the Gamechanger app. *Communication and Sport*, *8*(1), 72–91.

Schaefer, R. (2023). How NBA teams are using Sportradar's data collection platform this season. *Sports Business Journal*. Retrieved January 6, 2025, from https://www.sportsbusinessjournal.com/Articles/2023/10/27/nba-sportradar-synergy-sports

Schwarz, A. (2004). *The Numbers Game: Baseball's Lifelong Fascination with Statistics*. St. Martin's Press.

Schwarz, A. (2005). The great debate. *Baseball America*. Retrieved June 28, 2011, from http://www.baseballamerica.com/today/features/050107debate.html

Sheps, S. W. (2020). Corsi, Fenwick and Gramsci: How bloggers and advanced analytics are changing the National Hockey League. *International Review for the Sociology of Sport*, *55*(8), 1192–1211. https://doi.org/10.1177/1012690219869192

Shore, C., & Wright, S. (2018). Performance management and the audited self. In B. Ajana (Ed.), *Metric Culture: Ontologies of Self-Tracking Practices* (pp. 11–36). Emerald Publishing Limited.

Slater, M. (2023). FIFPRO agrees to develop system that will give players ability to control their own performance data. *The Athletic*. Retrieved December 3, 2024, from https://www.nytimes.com/athletic/4677583/2023/07/09/fifpro-agree-perfromance-data-system/

Smith, B. (2024). Examining the increasing correlation of home run rate and strikeout rate. Retrieved May 30, 2025, from https://rmsabr.org/blog/examining-the-increasing-correlation-of-home-run-rate-and-strikeout-rate/

Sports venue analytics: Leveraging data for fan engagement and operations. (2024). Sportsvenue technology. Retrieved January 6 from https://www.sportsvenue-technology.com/articles/sports-venue-analytics-leveraging-data-for-fan-engagement-and-operations

SSAC. (2024). *General Questions*. Retrieved November 20 from https://www.sloansportsconference.com/about/faqs

Stanley, J. (2024). Face recognition threatens to replace tickets, ID at sports events--and beyond. *ACLU*. Retrieved January 6, 2025, from https://www.aclu.org/news/privacy-technology/face-recognition-threatens-to-replace-tickets-id-at-sports-events-and-beyond

Swartz, T. (2024). Lebron James' biohacking routine reportedly costs $1.5M—Expert reveals how it'll affect his longevity. *MSN.com*. Retrieved January 3, from https://www.msn.com/en-us/sports/nba/lebron-james-biohacking-routine-reportedly-costs-15m-expert-says-it-will-help-him-live-a-long-time/ar-AA1tK0SU

Szymanski, S. (2020). Sport analytics: Science or alchemy? *Kinesiology Review*, *9*, 57–63. https://doi.org/10.1123/kr.2019-0066

Tenbarge, K. (2024). Facial recognition technology use at stadiums across the U.S. sparks protests. *NBCnews.com*. https://www.nbcnews.com/tech/security/facial-recognition-technology-use-stadiums-us-sparks-protests-rcna167410

Torre, P. S., & Haberstroh, T. (2014). New biometric tests invade the NBA. *ESPN.com*. Retrieved December 3, 2024, from https://www.espn.com/nba/story/_/id/11629773/new-nba-biometric-testing-less-michael-lewis-more-george-orwell

Tulle, E. (2016). Living by numbers: Media representations of sports stars' careers. *International Review for the Sociology of Sport*, *51*(3), 251–264. https://doi.org/10.1177/1012690214525157

Vishwarupe, V., Bedekar, M., Joshi, P. M., Pande, M., Pawar, V., & Shingote, P. (2022). Data analytics in the game of cricket: A novel paradigm. *Procedia Computer Science*, *204*, 937–944. https://doi.org/10.1016/j.procs.2022.08.114

Watanabe, N. M., Shapiro, S., & Drayer, J. (2021). Big data and analytics in sport management. *Journal of Sport Management*, *35*, 197–202. https://doi.org/10.1123/jsm.2021-0067

What is the 'Dunk Score?'. (2024). Retrieved November 21 from https://www.nba.com/news/what-is-the-dunk-score

Whelan, C. (2014). Surveillance, security, and sports mega events: Toward a research agenda on the organisation of security networks. *Surveillance & Society*, *11*(4), 392–404.

PART II

Emerging Technologies and Virtual Experiences

5

ARTIFICIAL INTELLIGENCE IN SPORTS AND RECREATION

Yoonki Chun and Philip Kang

Introduction

Artificial intelligence (AI) is a branch of computer science that focusses on developing machines that can emulate human-like behavior, performing tasks such as decision-making, problem-solving, and language processing (McCarthy, 2007). The transformative impact of AI on the sports and recreation industry is profound, reshaping sport management practices, enhancing athlete performance, improving fan engagement, and operational efficiencies (Pisaniello, 2024; Rahmani et al., 2024a). AI technologies, such as machine learning and predictive analytics, facilitate precise performance tracking and injury prevention by analyzing biomechanical data, thus optimizing training regimens and prolonging athletes' careers (Pisaniello, 2024; Rahmani et al., 2024a). Furthermore, AI enhances the fan experience through personalized content and immersive technologies, fostering deeper engagement and satisfaction (Lingappan et al., 2025). Affecting the efficiency of operating sport events, AI-driven systems, such as Automated Ball-Strike System (ABS) in baseball and Hawkeye in tennis, improve game officiating accuracy and streamline event management, ensuring fair play, and enhancing the spectator experience (Chen, 2024a; Rahmani et al., 2024a). For example, ABS tracks and collects pitched baseball's trajectory, speed, and relative position to the strike zone through the use of advanced technologies to determine whether the pitch passed the zone (Guéziec, 2002; Tseng et al., 2024). Leagues that adopted the system (e.g., Korea's KBO League) reported that it greatly increased the consistency and accuracy of pitch calls (Lee et al., 2024). Despite these advancements, challenges such as data privacy and ethical concerns remain, necessitating ongoing research to maximize AI's benefits in sports (Chen, 2024b; Pisaniello, 2024).

DOI: 10.4324/9781003588405-7

In consideration, the competing nature of AI between efficiency and ethical concerns presents significant opportunities alongside notable challenges (Adelakun et al., 2024), particularly for sport managers who must navigate ethical considerations and data privacy concerns. While AI can significantly enhance operational efficiency (e.g., game officiating, consumer relations), improve decision-making, and personalize experiences (Adelakun et al., 2024; Zaman, 2024), the integration of AI into sports also raises critical ethical issues, including transparency, accountability, and the potential for bias, which can undermine trust in AI systems (Adelakun et al., 2024; Gilbert & Gilbert, 2024). Moreover, the extensive data collection required for AI applications poses privacy risks, necessitating robust safeguards to protect sensitive information and comply with regulations (Gilbert & Gilbert, 2024). As AI evolves, sport management professionals must adopt comprehensive strategies that balance innovation with ethical standards and privacy protection to harness AI's benefits while mitigating risks that can affect athletes and sport consumers (Germani et al., 2024).

Aiming to provide resources in understanding key concepts and advancements in AI within the sports and recreation industry, this chapter introduces AI in athlete performance enhancement, democratization of AI technologies, AI-driven fan engagement strategies, and increasing focus on data privacy and ethics.

Background

The History of AI Development

Research in AI was founded at a workshop held on the campus of Dartmouth College in 1956 (Kaplan & Haenlein, 2019). After this Dartmouth Workshop, the programs that were developed amazed most people. Computers were solving algebra problems, proving geometry theorems, and even learning to speak English (Crevier, 1993; Russell & Norvig, 1995). In the past, assuming that machines could do those tasks was almost impossible. However, researchers were very positive. Attendees of the Dartmouth Workshop became the leaders of AI research for decades. Many predicted that machines would be as intelligent as humans within a single generation. The US government provided resources to achieve this vision (Newquist, 1994). Organizations like the Defense Advanced Research Projects Agency (DARPA) made significant investments in AI, which stimulated the establishment of AI labs at multiple universities in the United States and the United Kingdom during the late 1950s and early 1960s (Crevier, 1993). Despite the slight downfall of AI due to criticism and financial challenges in the 1970s, public interest in AI continued to rise, the number of researchers increased dramatically, and new ideas were explored in logic programming, commonsense reasoning and many other areas

(Crevier, 1993; Haigh, 2023). In the 1980s, the emergence of expert systems enabled computers to perform specific decision-making tasks (Russell & Norvig, 1995). For example, MYCIN, developed at Stanford in the 1970s, is a pioneering expert system for medical diagnosis (Shortliffe & Buchanan, 1975). It identified blood-borne infectious diseases and recommended antibiotics by processing patient data through if-then rules (Lea, 2023). Matching the accuracy of human experts, MYCIN showcased AI's potential in encoding specialized medical knowledge for clinical decisions. On the other hand, Dendral was created to tackle the complex task of identifying the structure of unknown organic molecules using mass spectrometry data (Lea, 2023). It assisted chemists by generating hypotheses about molecular structures based on spectral data and chemical constraints (Buchanan & Feigenbaum, 1981). Unlike general-purpose problem solvers, Dendral relied on a deep knowledge base of chemical rules and an inference mechanism tailored to its domain. Its success in automating a task that required expert-level chemical insight not only advanced scientific research but also laid the groundwork for future expert systems across various fields (Buchanan & Feigenbaum, 1981).

As computing power and data availability increased, the field of AI transitioned into machine learning techniques in the 1990s and 2000s, allowing AI systems to learn from vast amounts of data (i.e., big data) and improve the performance of machine learning on various tasks over time (e.g., image and video processing, text analysis, and speech recognition; LeCun et al., 2015).

AI-Powered Transformation in Sports

AI-driven innovation is transforming sports business models, strategy, and operations by leveraging data analytics to optimize decision-making and enhance organizational efficiency. Predictive models and AI-powered tools enable sports organizations to analyze market trends, streamline ticketing strategies, and maximize revenue through dynamic pricing (Glebova et al., 2024). These technologies support strategic planning by forecasting demand and improving resource allocation, allowing teams to adapt to competitive and financial pressures. Additionally, AI enhances operational workflows by automating tasks such as scheduling, logistics, and fan data management, reducing costs and creating opportunities for scalable business models (Galily, 2018). This integration of AI fosters a data-driven culture, positioning sports organizations to innovate and thrive in a rapidly evolving industry.

AI refers to the attempt to automate intellectual activities typically performed by humans (Du et al., 2023), which enables data-driven decision-making in sports. Due to the development of AI (e.g., machine learning), there have been key advancements in both on- and off-the-field areas in the sport industry. Wearable technology (e.g., Catapult GPS-accelerometer vests), powered by machine learning, now plays its role as an important data collection tool enhancing real-time data monitoring and producing predictive analytics. Similarly, the Internet of Things (IoT) (e.g., smart stadium and Intuit Dome) have transformed performance analysis, injury prevention, and safety (Kaur et al., 2025; Kovalchik, 2023; Li et al., 2016).

These interrelated technologies have transformed the sport industry. In motorsports, AI detects vehicle malfunctions, enhancing driver safety, while in soccer, it predicts and prevents injuries more effectively (Nalbant & Aydin, 2022; Van Eetvelde et al., 2021). AI also assesses injury risks and recovery times accurately, using data collected through wearable devices (Kakavas et al., 2022; Nicolella et al., 2018; Ráthonyi et al., 2018). On the other hand, sport data analytics, sourced by big data, wearable devices, and AI, enhances the understanding of team dynamics and training effectiveness (Li et al., 2016). Furthermore, AIoT (integration of AI with IoT) has introduced precision and predictive capabilities that outperform traditional methods.

The advancements in computer vision and virtual reality (VR) have further refined training experiences and engagement (Nalbant, 2021). Off the field, AI enhances fan experiences through personalized content and immersive technologies, while improving operational efficiencies in sports organizations via predictive models and automation (Chen, 2024a; Glebova et al., 2024). Understanding how these technologies develop with AI is crucial for grasping its crucial impact on sport management.

Topical Exploration and Key Concepts

AI and Athlete Performance Enhancement

AI technologies are dramatically affecting the sports industry by transforming how athletes train, recover, and perform. Such integration of AI allows for the collection and analysis of complex big data, providing data-driven insights that can assist decision-making related to athlete development and competency (Claudino et al., 2019). This section highlights significant advancements in athlete performance that actively use AI technology: wearable technology and biometric monitoring, performance analytics, and personalized training programs.

Wearable Technology and Biometric Monitoring

AI in Collecting Real-Time Biometrics

Wearable technology with AI capabilities has become the new standard in modern sports training. Various sensors can collect real-time data on various physiological and biomechanical parameters, including heart rate, muscle activity, movement patterns, and stress levels (Afsar et al., 2023; Duan et al., 2024). Then, AI algorithms process these data to monitor athletes' performance and health status with a precision level that exceeds human measurement. Advanced learning models such as convolutional neural networks (CNNs) and other deep learning architectures have been implemented to analyze datasets collected from wearable devices (Duan et al., 2024). These models can identify small changes in biometric data that can be used as indicators for fatigue or increased risk of injury (Claudino et al., 2019).

Applications in Training Optimization and Injury Prevention

Playing a vital role in optimizing training programs and preventing injuries, AI-driven biometric monitoring analyzes patterns in training loads, biomechanical markers, and physiological indicators. Through this, AI models can predict potential injuries prior to their occurrence among athletes, allowing proactive interventions (Claudino et al., 2019; Malone et al., 2017). For example, using support vector machines (SVMs) and decision tree classifiers has proven its effectiveness in forecasting potential risk factors using data from wearable sensors (Rossi et al., 2018; Ruddy et al., 2018).

Moreover, AI provides an enhanced personalization of training regimens tailored to individual athletes. Data collected from wearables can be customized to meet individual athlete needs, considering factors such as age, gender, genetics, training, and injury history (Duan et al., 2024). Such a personalized approach improves performance-related indices while minimizing potential risks for injury. Integrating multimodal data, including video analysis, sensor data, and physiological monitoring, improves the accuracy and comprehensiveness of performance analysis through multimodal data fusion technology (Duan et al., 2024).

Performance Analytics

AI-Driven Data Analysis for Strategies and Tactics

AI has made significant shifts in how performance analytics is presented by enabling the processing of complex and large datasets that can be readily used to uncover previously unattainable insights. Machine learning models such as artificial neural networks (ANNs), decision tree classifiers, and SVMs are implemented to further enhance strategic decision-making in sports (Galekwa et al., 2024). These techniques analyze technical and tactical aspects, providing coaches with data-driven insights to elevate team performance.

In sports biomechanics, AI-enhanced motion capture systems are used to accurately assess athlete's subtle movements. These systems utilize advanced algorithms like CNNs to process and analyze motion data collected through high-precision sensors and videos. For example, research indicates that these AI-enhanced motion recognition can achieve an intersection over union (IoU) of 0.9 and an F1 score of 0.91, indicating superior performance compared to other traditional mechanisms (Liu et al., 2024). Additionally, by integrating computer vision techniques, systems can identify erroneous movements, further enhancing training accuracy to over 90% (Xie & Xu, 2024).

Machine Learning Models Predicting Performance Outcomes

Machine learning models have demonstrated superior predictive power compared to traditional statistics in forecasting performance-related outcomes in sports. For instance, AI methods have been actively utilized in soccer to measure injury risk and predict athlete performance using training and biomechanical data (Baratela et al., 2024; Musat et al., 2024). Relatedly, Rossi et al. (2018) demonstrated the effectiveness of machine learning in predicting injuries based on training data, while Nowak et al. (2024) reported a relative error of 2.3% in predicting extreme performance indices using linear models. These predictive models can support coaches and related professionals to make informed decisions regarding team selection, training intensity, and game strategies, contributing to a competitive edge within sports.

Personalized Training Programs

AI algorithms analyze individual athletes to create customized training programs, considering their strengths, weaknesses, and goals (Gabbett, 2010). AI facilitates the dynamic adaptation of training programs, and machine learning models can intervene and adjust training loads in real-time based on athletes' current condition, physiological responses, and performance metrics (Claudino et al., 2019). This approach can maximize training potential and efficiency while reducing the risk of experiencing overtraining and injury.

Furthermore, AI offers novel paths for personalized health enhancement, such as nutrition plans integrated with training regimens (Puce et al., 2024). Yet, despite its benefits, the effectiveness and accuracy of AI-generated advice toward athlete progression should be rigorously tested and validated to ensure reliability. Hence, future research may focus on exploring the specific mechanisms of nutritional factors, optimal nutrient ratios, and the timing of the interventions to take full advantage of AI's potential concerning assisting nutrition in sports (Duan et al., 2024).

In tandem, AI technologies significantly enhance athlete performance by providing advanced tools for biometric monitoring, performance analytics, and personalized training. Integrating AI in sports science offers accurate, data-driven understandings that support injury prevention, strategic decision-making, and personalized athlete development. Yet, challenges remain, such as the need for more extensive and more diverse datasets, integrating multimodal data sources, and addressing individual differences among athletes.

AI in Sport Management Practices

AI-Enhanced Sport Management: Efficiency, Automation, and Predictive Power

This section explores how AI influences sport organization's management practices. The sports industry is increasingly leveraging technology to enhance operational efficiency (Rahmani et al., 2024b). AI plays a crucial role in sport management by improving decision-making, reducing labor costs, and enhancing service quality through automation of tasks like ticket sales and customer service. For instance, AI chatbots can handle customer inquiries, boosting fan engagement (Chu, 2024). Additionally, AI's predictive models are invaluable for facility managers. They allow for continuous monitoring of sports facilities and equipment, helping to identify potential hazards and schedule maintenance to prevent breakdowns, extend asset lifespan, and lower repair costs (Chin et al., 2022).

Predictive models can effectively manage energy usage in stadiums by adjusting electrical systems and climate control according to real-time data, which leads to significant savings in operational costs (Elnour et al., 2021). Additionally, when predictive models are combined with robotics, they can improve manufacturing processes, resulting in more efficient and sustainable production of sports equipment (Adebayo, 2024). Furthermore, predictive models assist business managers in optimizing ticket pricing, forecasting demand, and enhancing marketing strategies. This ultimately maximizes revenue and enriches the fan experience through personalized content and recommendations powered by AI (Glebova, 2024).

AI's Impact on Leadership and Ethics in Sports Management

These changes, leveraged by AI, influence how leaders make organizational decisions (Korzynski, 2023; Shrestha et al., 2019). By providing real-time data and actionable insights (Okech, 2024), AI has transformed organizational roles and responsibilities by automating repetitive tasks, which enables leaders to focus on strategic and creative initiatives rather than devoting their time to those time- and energy-consuming tasks (Wang, 2021). Moreover, as AI becomes an integral part of the decision-making processes (Abu Nasser et al., 2024), the importance of ethical leadership has grown significantly. Especially, addressing bias in AI algorithms and maintaining a human-centric approach is essential (Rahmani et al., 2024b). AI algorithms trained on historical data may carry existing biases. Thus, when leaders use these algorithms for decision-making, it can result in biased outcomes.

For example, in recruitment practices, they might discriminate against underrepresented ethnic groups. They may favor specific demographics based on biased historical data, resulting in unintended discrimination and a lack of diversity (Hughes, 2023). To mitigate this risk, achieving fairness and transparency in AI models is crucial (Khan, 2023). Several methods can be applied to achieve this goal. For example, regular audits, diverse training datasets, and ongoing monitoring can help reduce bias and increase fairness in the results of applied AI (Dindorf et al., 2022). Additionally, upskilling leaders to balance technological expertise with human qualities such as intuition and emotional intelligence is critical for the future (Shah, 2024).

A fundamental change in leadership training programs is essential to equip leaders for a world influenced by AI. While AI delivers challenges and opportunities, its successful implementation depends on leaders' ability to adapt, foster human—AI interaction, and ensure that AI is utilized as assistance rather than replacing human capacity. Thus, future research should further explore AI's complicated impact on leadership, refining existing insights to create a more effective and ethically sound integration of AI in sports management. Despite its potential, integrating AI into sports also entails several challenges. One significant obstacle is the high cost of AI implementation, which can be a barrier for smaller organizations due to substantial upfront investments in software, hardware, and infrastructure. Additionally, the complexity of AI technologies requires specialized skills and knowledge, making it challenging for some organizations to adopt and maintain AI systems effectively (Al-Jenaibi & Bulhoon, 2024).

AI in Fan Engagement and Marketing

This section explores how AI impacts fan engagement and strategic marketing in the sports industry. Fan engagement is crucial for sports clubs and organizations to achieve continuous success (Mazza & Russo, 2023). In recent years, the sports industry has been actively introducing high-tech solutions worldwide, such as analytics (Rangasamy et al., 2020). For instance, goal-line technology, wearable fitness trackers, and applied AI analytics demonstrate how technology has become essential to sports. These innovative tools have enhanced the spectators' experience, and teams and athletes can develop their performance with invaluable data (Ramkumar et al., 2022).

Revolutionizing Fan Engagement and Sports Marketing

The integration of 5G mobile communication, AI-based data analytics, and the Internet of Things (IoT) in sports (Bhatia, 2020; Gowda et al., 2018) has driven the growing demand for advanced sensor technologies. AI transforms fan engagement and sport marketing by providing personalized experience, real-time interaction, and data-driven decision-making (Peranzo, 2024; Rafalski, 2024). For example, sports organizations leverage AI to enhance consumers' brand loyalty, optimize marketing strategies, and drive maximum revenue using AI-driven chatbots and immersive augmented reality experiences (Barraud, 2024).

The convergence of AI and sports marketing has created a transformative force reshaping how businesses engage with sports consumers. AI is increasingly applied across diverse fields to enhance operational efficiency, decision-making, and customer engagement (Babatunde, 2024). In sports marketing, AI's integration has transformed traditional practices by enabling real-time data analysis, predictive modeling, and personalized communication, which fundamentally impact interactions among fans, sponsors, and other stakeholders (Beal et al., 2019). Historically, the field has relied heavily on branding, fan engagement, and media partnerships (Ajeigbe, 2021); however, the rise of digital platforms and AI-driven technologies now offers more dynamic and interactive ways to reach and engage audiences.

Due to this phenomenon, AI-powered algorithms increase fans' engagement by analyzing their data to deliver personalized content, promotions, and experiences (Sarkar et al., 2024). AI algorithms can recommend tailored merchandise, ticket offers, and exclusive content by evaluating fan preferences and engagement histories. This AI-driven personalized function can be extended to customer service as well. AI chatbots and virtual assistants provide instant, customized responses, facilitate ticket purchases, and offer comprehensive game-day information, which maximizes fans' engagement toward the game (El-Maghrabi & Sharif, 2022). Furthermore, using real-time sentiment analysis and emotional AI, sports organizations can gauge audience emotions during live events by monitoring social media and digital platforms. This capability enables dynamic adjustments to marketing campaigns and live experiences and creates more resonant and impactful interactions. In addition, predictive models forecast fans' behavior and engagement patterns, which assist sport organizations to proactively design effective engagement strategies and campaigns targeting fans (El-Maghrabi & Sharif, 2022).

AI-Powered Personalization, Immersion, and Business Growth

AI's role in sports marketing has led to significant business advancements by analyzing large data sets, predicting consumer behavior, and providing personalized experiences. Additionally, AI-powered customer segmentation refines marketing outreach, ensuring that promotions reach the most receptive audiences (Thilagavathi, 2024). Applications such as machine learning, natural language processing, and applied AI analytics help sport marketers to create highly personalized campaigns, enhance customer engagement, and optimize revenue streams through targeted advertising and sponsorship opportunities. For instance, IBM's Watson and various fan engagement tools analyze data from social media and other channels to track fans' interactions and create customized content that aligns with individual preferences (Bin & Xu, 2021; Swant, 2024).

Moreover, AI-enhanced platforms are critical in optimizing sports marketing operations by automating routine tasks, analyzing massive datasets to uncover consumer insights, and facilitating real-time decision-making (Chintalapati & Pandey, 2021). Sports teams can now use AI to predict ticket sales patterns, optimize dynamic pricing, and develop highly targeted digital marketing campaigns that appeal to specific fan demographics (Chin et al., 2022). Since we live in an environment where consumers expect tailored and interactive engagement, delivering personalized content, such as video highlights and merchandising offers, has become essential to enhance customer service satisfaction (Babatunde, 2024).

Integrating AI and augmented reality (AR) and VR further enriches the fan experience by offering immersive opportunities such as virtual meet and greets, AR-enhanced game broadcasts, and interactive stadium experiences. These features allow fans to experience the game from a player's perspective and enhance fans' understanding of the sport beyond traditional viewing (El-Maghrabi & Sharif, 2022). Furthermore, combining AI and blockchain technology (e.g., NFTs) ensures secure and transparent transactions for ticket sales, merchandise, and digital collectibles. At the same time, AI-driven loyalty programs reward fans based on engagement and participation (El-Maghrabi & Sharif, 2022). Throughout these technological advancements, ethical considerations remain paramount to ensure that fan data is managed responsibly and with proper consent (El-Maghrabi & Sharif, 2022).

Lastly, mobile applications and social media platforms enhance fan engagement by updating real-time data and providing gamified features and exclusive content. These digital tools enhance the connections between fans and sports organizations, eventually offering innovative revenue streams and amplifying brand loyalty through direct and interactive communication. Thus, endorsing personalization, gamification, and reward systems within these platforms is essential to maintaining fans' engagement (Bayindir, 2024).

Challenges

Despite its advantages, AI in fan engagement and marketing presents challenges. AI has raised serious privacy issues that have hindered technology adoption (Fang & Tse, 2022). Students from technology and science-focused schools displayed a higher degree of concern, which suggests their more profound understanding of potential privacy implications (Irfan et al., 2023). Also, using AI-driven personalization in digital marketing and other fields brings about ethical challenges related to privacy, bias, manipulation, and societal impacts (Karami et al., 2024). Also, collecting user data without explicit permission and using it to make inferences raises significant moral problems (Sarioguz & Miser, 2024). It is especially imperative for adolescents, who are key users of online platforms, to develop a deep understanding of these practices to protect their privacy (Khan et al., 2024). Thus, sport organizations must ensure compliance with data protection regulations and maintain transparency in AI-driven personalization efforts.

Ethical Issues and Data Privacy in AI Applications in Sport

As previously mentioned, integrating AI technologies into sport has revolutionized industry performance, data analytics, and decision-making. Its ability to process extensive data provides unprecedented insights into athlete performance, injury prevention, and fan engagement. Yet, this rapid adoption of AI accompanies ethical concerns that must be addressed to uphold the integrity of sports, protect athlete well-being, and ensure equal competition (Trail et al., 2024).

One of the foremost concerns is data privacy. AI systems are operated by collecting and analyzing large amounts of personal and sensitive data, so they risk potential misuse or unauthorized access (Resnik, 2015). Ethical guidelines and legal frameworks are vital to navigating these challenges, but their implementation in sports presents difficulties (Nguyen et al., 2023). This section explores the ethical issues and data privacy concerns associated with AI applications in sports, emphasizing the need for robust ethical standards and practices.

Ethical Issues in AI Applications

AI algorithms can unconsciously extend biases in their training data, leading to unfair treatment of particular athletes or groups. Bolukbasi et al. (2016) asserted how AI systems could reinforce gender stereotypes through biased word embeddings. Such biases may affect talent scouts, performance evaluations, and decision-making processes in sports, undermining fairness and equity.

Moreover, disparities in access to AI-related technologies between teams with varying available recourses can also exacerbate existing inequalities (Pfleegor et al., 2019). Wealthier organizations may leverage advanced AI tools to improve performance, potentially giving them an unfair competitive advantage over less resourceful organizations. Such inequality challenges the integrity of competitions and raises ethical questions about providing equal opportunity. Hence, ethical principles demand that all competitors have fair access to performance-enhancing technologies to maintain the spirit of sport (Trail et al., 2024).

Extending the current concerns, the data generated by athletes during training sessions and competitions raises questions concerning ownership and its control. While ethics concerning ownership is a topic of interest in various disciplines (Pfleegor et al., 2019), using athletes' data without appropriate acknowledgment or compensation can become a rising issue as AI becomes more relevant. While athletes may have little interest in or have limited control over such data, it bears the potential of exploitation.

Organizations might consider commercializing athlete data for profit, such as selling it to third parties or using it for marketing purposes without proper authorization (Trail et al., 2024). For example, video games that utilize actual athlete data (e.g., EA Sports FC, NBA2K) may advance to using real-time data generated with advanced AI technologies. Such potential for exploitation further raises ethical concerns regarding intellectual property rights and fair compensation.

Recommendations

Organizations should adopt ethical frameworks that guide AI implementation, emphasizing key principles like honesty, integrity, and accountability. Such guidelines can help promote responsible innovation that benefits all stakeholders. Further, athletes, coaches, administrators, and technologists should be involved collectively in developing ethical policies to ensure diverse perspectives (Pfleegor et al., 2019). Such collaborative efforts can foster a shared understanding and commitment to these ethical practices concerning AI.

Transparent disclosure of AI uses and data handling can help build trust within organizations. They should also provide clear information to their members and external stakeholders, facilitating informed decision-making. Additionally, providing ethical training for researchers and practitioners underscores the importance of integrity and responsible conduct (Nguyen et al., 2023). Continuous education supports adherence to ethical standards.

Implementing robust consent procedures ensures athletes are fully informed and agree to the collection of data and its usage through AI. Respecting their autonomy can further enhance ethical compliance. Investing in advanced cybersecurity measures can protect sensitive data from unauthorized access (Pion-Tonachini et al., 2021). Implementing secure systems within organizations is essential for maintaining privacy and meeting legal obligations.

Overall, the application of AI in sports presents both great opportunities and significant challenges concerning ethics. Addressing issues of fairness, autonomy, data privacy, and legal compliance is essential to fully harness the potential of AI while safeguarding the integrity of sport and the welfare of its stakeholders. By developing comprehensive guidelines, promoting transparency, and fostering collaboration among stakeholders, the sport industry can continue to navigate the complex nature of AI integration responsibly. Continued communication and commitment to these ethical principles can further enhance, rather than detract from, the spirit of competition and the values that underpin athletic endeavors.

Practice in Action

The US Open Tennis Championship, held annually in Flushing, NY, is one of the most highly attended sporting events globally, inviting nearly one million visitors. Moreover, the event extends beyond its attendees on-site, with over 15 million tennis fans following the action through the US Open website and mobile app. This expansive engagement is credited from the long-standing partnership between the United States Tennis Association (USTA) and IBM Consulting, which has delivered a state-of-the-art digital experience for over 30 years.

In 2024, the USTA and IBM recently implemented innovative generative AI models to enhance the digital experience of the US Open tournament. Through IBM's Watsonx platform, they focused on utilizing match data and transforming them into insightful content for tennis fans. Key features developed include AI-generated match reports and commentaries for highlight content, providing fans with quick and comprehensive updates on the actions at the tournament (Goldman, 2023).

The US Open is an extensive, two-week event with hundreds of matches across 22 courts. Managing and handling the vast amount of data generated through these matches greatly benefits from generative AI-powered features using the Granite foundation models developed by IBM Consulting. These large language models, trained with high-quality tennis data, generate match statistics into coherent summaries and commentary, offering fans a richer and more comprehensive insight into the tournament.

The integration of AI extends beyond enhancing fan engagement. By automating important workflows, AI models significantly improve the efficiency of the USTA's editorial team, dramatically increasing the matches being covered. The ongoing operation of these models is monitored using Watsonx governance tools, maintaining compliance and performance. IBM's approach to innovation stems from collaboration and agility. The IBM Garage Methodology facilitates co-creation with the USTA, allowing for the rapid development and implementation of new digital capabilities. This flexible platform processes structured and unstructured data, integrating technology from various sources to facilitate a seamless digital experience. The US Open generates over seven million data points, including player statistics and match-specific details like serve speed and rally count. With this structured data, IBM's data management operation integrates unstructured media commentary for the fans' comprehensive view of each match. Here, using IBM Watsonx data, a purpose-built data store, supports such a massive data integration.

Overall, the partnership between USTA and IBM exemplifies how AI technology can advance and transform traditional sports by creating dynamic and engaging experiences for fans through fabricated data. As the digital landscape in AI continues to move forward, such collaboration is expected to push the boundaries of innovation in the sport industry, offering new possibilities for its future.

Trends and Current Issues

Increasing Focus on Data Privacy and Ethics

Reflecting the rapid growth of AI and big data, organizations quickly acknowledge the importance of data privacy and implement ethical regulations. Unsurprisingly, 90% of the world's data has been generated in the past three years (Patel, 2024). While the tremendous amount of data opens up new opportunities and growth, it has also been noted that 81% of Americans are worried about potential threats and misuse regarding data analytics (McClain et al., 2023). Such concerns largely stem from the risk of invasion of privacy, algorithmic bias, and lack of transparency (Patel, 2024). Reacting to public concerns, recent advancements have introduced comprehensive data protection laws, providing insight and guidance for implementing data analytics across various practices.

The General Data Protection Regulation (GDPR), California Consumer Privacy Act (CCPA), and California Privacy Rights Act (CPRA) are landmark data protection laws that have reshaped how organizations handle personal data, with significant implications for the sport industry. These regulations aim to improve data privacy and give individuals better control over their personal information. The GDPR, implemented in 2018, affects organizations that process EU citizens' private data, providing comprehensive protection (Peukert et al., 2022). The CCPA, also implemented in 2018 alongside CPRA, which started in 2023, reflects the state of California's effort to address increasing concerns regarding data privacy, which applies to businesses that attempt to collect data from its residents (Layton & Elaluf-Calderwood, 2019). Such recent introductions of new regulations pose strict requirements for collecting, processing, and sharing individual data while penalizing noncompliance. Hence, the sport industry, which also increasingly relies on data analytics for decision-making, fan engagement, and athlete performance tracking, must be attentive to these increasing demands.

Furthermore, GDPR dramatically limits the size and purpose of data usage and strictly requires user consent (Zaman et al., 2019). Thus, when using wearable devices to assess athletes' performance, for instance, teams should be aware of GDPR's rules and comply with them for data protection and transparency (Ebbers, 2023). On a similar note, CCPA and CPRA allow California residents to know the type of data being collected, to access the data, and to request its deletion, which sports organizations must conform to when performing data analytics for fans and athletes (Park, 2019).

These new regulations also apply to the sport industry's digital infrastructure, such as websites and mobile apps, which often use cookies and tracking technologies for user experience and target marketing (Pantelic et al., 2022). GDPR has led to the widespread adoption of cookie consent notices, requiring active consent by users for data collection, which is now standard practice in many regions (Degeling et al., 2018). Recent laws and regulations like GDPR, CCPA, and CPRA are introducing a new era of data protection that requires the sport industry to reevaluate its data utilization strategies. While these regulations may present additional challenges and barriers, they also allow organizations to further build trust with fans and athletes through data privacy and transparency. As the sport industry continues to advance with the increase of AI and data analytics, compliance with new and upcoming laws will remain a focal area.

Democratization of AI Technologies

The democratization of AI makes AI tools and technologies accessible to a wide range of people and organizations (Johnson, 2020). This phenomenon is driven by the availability of training data, cost-effective tools, and efforts to promote widespread adoption (Costa et al., 2024; Himmelreich, 2022). A clear example of this trend is that generative AI models are increasingly being adopted by businesses, government agencies, and countries to meet their goals and strategies.

Accessibility of AI to Smaller Organizations

The growing access to AI is breaking down the barriers that kept powerful technology in the hands of big companies. Now, more individuals and smaller businesses can utilize these advanced tools. Cloud-based AI services, such as Amazon Web Services (AWS), Google Cloud, and Microsoft Azure, have been pivotal in this shift by offering scalable, on-demand resources. These services enable small and medium-sized enterprises and independent developers to leverage AI functionalities (Bermudez et al., 2013; Madhuri & Sowjanya, 2016; Saraswat & Tripathi, 2020; Ucuz, 2020).

Improvements in education support this trend, as universities and platforms like Coursera and edX offer certifications in AI and machine learning, democratizing access and fostering diverse AI practitioners (Aparicio et al., 2014, 2019; Baturay, 2015). Additionally, government policies embrace AI to boost the economy and enhance public services. Initiatives like the European Union's AI strategy (EU, 2024) and the United States AI for the American People (White House, 2023) aim to promote research, education, and ethical AI deployment, ensuring equitable access to its benefits (Ernst et al., 2019).

Open-Source AI Tools and Platforms

Open-source accessibility is essential for democratizing AI, enabling global collaboration and innovation with tools like TensorFlow, PyTorch, and scikit-learn (Costa et al., 2024). This commitment aligns with the demand for AI systems to embody attributes such as portability, explainability, credibility, and fairness, which require advancements in research, technology, and regulations (Ahmed et al., 2020). Large-scale distributed AI systems enhance collaboration and learning beyond traditional federated learning, allowing for better adaptation to various tasks (Nguyen et al., 2023). Consequently, open-source tools are widely accessible and adaptable to diverse user needs and environments.

Equitable Competition Across Sectors

The democratization of AI technologies allows smaller companies to compete with traditional leaders in the market in various industries. By leveraging affordable cloud-based tools and open-source platforms, startups and independent developers can innovate without significant financial burden (Bermudez et al., 2013). Government actions also play an important role in promoting fairness by creating rules that protect information and privacy at both national and international levels. (EU, 2024; White House, 2023). Moreover, establishing public data trusts can help manage the data to train AI algorithms and ensure that economic values created by AI are shared impartially, which reduces the negative impacts and fosters a sense of fairness (Chan et al., 2023).

Enhanced Innovation Through Diverse Participation

The democratization of AI encourages innovation by expanding the user pool, which can produce new and creative ideas. Open-source tools and accessible educational resources also allow individuals and organizations globally to try out new ideas and create innovative solutions (Aparicio et al., 2019; Costa et al., 2024). Conversely, this increased participation raises social and ethical concerns. It emphasizes transparency, privacy, and fairness in AI applications, especially in data-intensive fields (e.g., energy management; Marinakis, 2021). Addressing these issues needs democratic governance initiatives emphasizing collective responsibility in AI decision-making (Garvey, 2018). Ultimately, with appropriate management of AI democratization, the integration of the accessibility of AI with users' novel ideas or perspectives not only fuels creativity but also yields innovations responsive to a wide array of societal needs and values (Costa et al., 2024; Himmelreich, 2022).

AI-Driven Fan Engagement Strategies

Shift Toward Digital Fan Experiences

AI significantly contributes to sports management and administration, influencing gameplay and fan engagement (Xu & Baghaei, 2025). AI-driven platforms utilize real-time fan data, such as preferences and viewing habits, to offer personalized content, including highlight reels and exclusive interviews (Buchheit et al., 2012; Santos-Gago et al., 2019; Tyler et al., 2022). This tailored approach enhances fan interaction with teams and athletes, leading to a more enjoyable experience. Furthermore, the integration of AI-powered immersive technologies like AR and VR enriches fan experiences (LeCun et al., 2015). AI-driven AR applications provide real-time player stats, alternative camera angles, and virtual meet-and-greets with athletes. Additionally, AI enhances accessibility for fans with disabilities through personalized features like closed captioning and audio descriptions, ensuring inclusivity in digital sports experiences (Tussyadiah & Pesonen, 2016).

Use of AI to Enhance Remote and Virtual Engagement

The effort to embrace AI-powered personalization exhibits the sport team's eagerness to foster strong connections with their fan base. Fans feel valued and understood, leading to increased engagement and potentially higher revenue (Xu & Baghaei, 2025). Especially chatbots and virtual assistants represent one of the prominent areas where sport teams try to provide a more personally tailored experience by leaning on AI technology, which can enhance fan engagement significantly (Pratisto et al., 2022). These tools changed the way they traditionally communicated with fans. For example, sport teams can utilize chatbots to offer live game updates or ticket availability. Virtual assistants can take this further by analyzing individual preferences and viewing history to provide custom recommendations. This real-time support and personalization strengthen the connection between teams and their fans, ultimately leading to greater engagement and revenue generation. Therefore, the impact of AI on the fan experience is substantial (Huang et al., 2021).

New Avenues for Revenue Generation

Integrating AI into digital fan engagement significantly boosts revenue potential. AI-driven personalized marketing optimizes merchandising and ticketing strategies, maximizing revenue through targeted promotions (Huang et al., 2021). By analyzing demand, demographics, and purchasing behaviors, AI algorithms create dynamic pricing models that enhance income (Hulin et al., 2016). AI also enables targeted ads using fan data, resulting in more relevant promotions and increased sales for sponsors. Additionally, AI-powered ecommerce integrations facilitate seamless in-app purchases, allowing fans to buy merchandise while watching live events. Offering customized experiences through AI enhances fan engagement and drives financial growth for sports organizations (Goes et al., 2019). Ultimately, AI integration boosts operational efficiency and aids managers in exploring new revenue avenues, positioning businesses for success in a competitive market.

Challenges in Keeping Fans Connected in a Digital Landscape

AI-driven fan engagement faces several challenges. The primary concern is data privacy and security, as fans may feel uncomfortable with extensive data collection without strong protections in place (Huang et al., 2021). Establishing clear regulations is essential for building trust between sports organizations and fans. Another challenge is algorithmic bias, which can lead to exclusionary outcomes; thus, it's important for engagement tools to be regularly refined for fairness and inclusivity. Additionally, excessive reliance on AI can reduce the authenticity of fan interactions, as automated responses often lack a human touch. Therefore, while pursuing efficiency with AI, it's crucial to maintain genuine interactions. Addressing these issues will help sports organizations create sustainable, interactive, and valuable engagement for fans and stakeholders.

Summary

This chapter examined the evolving role of AI in sports, discussing its historical roots and advancing through its impact on athlete performance, sport management, fan engagement, and ethical practice. We explored how AI-powered technologies—like wearable devices, predictive analytics, and immersive platforms—are reshaping training, injury prevention, operational efficiency, and personalized fan experiences. The chapter also addressed critical ethical and legal challenges, including data privacy and algorithmic bias, and emphasized the importance of transparent, responsible AI integration. Finally, the democratization of AI was discussed as a driving force expanding access and innovation across the industry. As AI continues to evolve, sport professionals must balance technological advancement with ethical leadership and inclusive practice.

Key Resources

- **IBM Watson X**: https://www.ibm.com/watsonx
- **IBM AI Ethics**: https://www.ibm.com/think/topics/ai-ethics
- **Athletica**: https://athletica.ai
- **Humango**: https://humango.ai/

- **VueMotion**: https://www.vuemotion.com/
- **AiScout**: https://www.wired.com/story/ai-football-soccer-scouting/

References

Abu Nasser, B. S., Saleh, M. M., & Abu-Naser, S. S. (2024). Implications and applications of artificial intelligence in the legal domain. *International Journal of Academic Information Systems Research*, *7*(12), 18–25.

Adebayo, R. A., Obiuto, N. C., Olajiga, O. K., & Festus-Ikhuoria, I. C. (2024). AI-enhanced manufacturing robotics: A review of applications and trends. *World Journal of Advanced Research and Reviews*, *21*(3), 2060–2072.

Adelakun, B. O., Majekodunmi, T. G., & Akintoye, O. S. (2024). AI and ethical accounting: Navigating challenges and opportunities. *International Journal of Advanced Economics*, *6*(6), 224–241.

Afsar, M. M., Saqib, S., Aladfaj, M., Alatiyyah, M. H., Alnowaiser, K., Aljuaid, H., Jalal, A., & Park, J. (2023). Body-worn sensors for recognizing physical sports activities in Exergaming via deep learning model. *IEEE Access*, *11*, 12460–12473.

Ahmed, S., Mula, R. S., & Dhavala, S. S. (2020). A framework for democratizing AI. *arXiv preprint arXiv:2001.00818*.

Ajeigbe, I. Y. (2021). Influence of sports marketing strategies in the promotion of organization products and services in Ilorin Metropolis. *Indonesian Journal of Sport Management*, *1*(2), 128–137. https://doi.org/10.31949/ijsm.v1i2.1311

Al-Jenaibi, B., & Bulhoon, A. (2024). Investigating adoption determinants, obstacles, and interventions for AI implementation in Emirati media organizations. *South Eastern European Journal of Public Health*, *25*, 1901–1909.

Aparicio, M., Bacao, F., & Oliveira, T. (2014, May). MOOC's business models: Turning black swans into gray swans. In J. Smith & A. Brown (Eds.), *Proceedings of the International Conference on Information Systems and Design of Communication* (pp. 45–49). Association for Computing Machinery.

Aparicio, M., Oliveira, T., Bacao, F., & Painho, M. (2019). Gamification: A key determinant of massive open online course (MOOC) success. *Information & Management*, *56*(1), 39–54.

Babatunde, S. O. (2024). The role of AI in marketing personalization: A theoretical exploration of consumer engagement strategies. *International Journal of Management & Entrepreneurship Research*, *6*(3), 936–949. https://doi.org/10.51594/ijmer.v6i3.964

Baratela, E. A., Xavier, F. J., Peron, T., Ribeiro Villas-Boas, P., & Rodrigues, F. A. (2024). Predicting soccer matches with complex networks and machine learning. *Journal of Complex Networks*, *12*(6), cnae043.

Barraud, C. (2024, December 20). How artificial intelligence is transforming sports marketing. *LinkedIn*. https://www.linkedin.com/pulse/how-artificial-intelligence-transforming-sports-cyrille-barraud-hcuze/

Baturay, M. H. (2015). An overview of the world of MOOCs. *Procedia-Social and Behavioral Sciences*, *174*, 427–433.

Bayindir, M. (2024). The role of mobile applications and social media in increasing fan loyalty. *Turkish Journal of Sports Informatics and Technology*, *1*(1), 1–13.

Beal, R., Norman, T. J., & Ramchurn, S. D. (2019). Artificial intelligence for team sports: A survey. *The Knowledge Engineering Review*, *34*, e28.

Bermudez, I., Traverso, S., Mellia, M., & Munafo, M. (2013, April). Exploring the cloud from passive measurements: The Amazon AWS case. In J. Smith & A. Brown (Eds.), *2013 Proceedings IEEE INFOCOM* (pp. 230–234). IEEE.

Bhatia, M. (2020). Iot-inspired framework for athlete performance assessment in smart sport industry. *IEEE Internet of Things Journal*, *8*(12), 9523–9530.

Bin, L., & Xu, X. (2021). Application of artificial intelligence in basketball sport. *Journal of Education Health and Sport*, *11*(7), 54–67.

Bolukbasi, T., Chang, K.-W., Zou, J. Y., Saligrama, V., & Kalai, A. T. (2016). Man is to computer programmer as woman is to homemaker? Debiasing word embeddings. *Advances in Neural Information Processing Systems*, *29*, 1–9.

Buchanan, B. G., & Feigenbaum, E. A. (1981). Dendral and Meta-Dendral. In B. L. Webber & N. J. Nilsson (Eds.), *Readings in Artificial Intelligence* (pp. 313–322). Morgan Kaufmann.

Buchheit, M., Simpson, M. B., Al Haddad, H., Bourdon, P. C., & Mendez-Villanueva, A. (2012). Monitoring changes in physical performance with heart rate measures in young soccer players. *European Journal of Applied Physiology*, *112*, 711–723.

Chan, A., Bradley, H., & Rajkumar, N. (2023, August). Reclaiming the digital commons: A public data trust for training data. In F. Rossi, S. Das, J. Davis, K. Firth-Butterfield, & A. J. London (Eds.), *Proceedings of the 2023 AAAI/ACM Conference on AI, Ethics, and Society* (pp. 855–868). Association for Computing Machinery.

Chen, J. (2024a). The application and development of artificial intelligence and high technology in sports event. *Highlights in Business, Economics and Management*, *30*, 247–255.

Chen, S. (2024b). *Real-time Data Analysis in Sports: Unleashing the Power of Analytics*. Retrieved February 8, 2025, from https://www.preciser.io/blog/real-time-data-analysis-in-sports-unleashing-the-power-of-analytics

Chin, J., Do, C., & Kim, M. (2022). How to increase sport facility users' intention to use AI fitness services: Based on the technology adoption model. *International Journal of Environmental Research and Public Health*, *19*(21), 14453.

Chintalapati, S., & Pandey, S. K. (2021). Artificial intelligence in marketing: A systematic literature review. *International Journal of Market Research*, *64*(1), 38–68.

Chu, Y. R., Wang, S., Wang, X., & Wang, J. (2024). Emerging trends in sports and artificial intelligence: A scientometric analysis in citespace. *Journal of Electrical System*, *20*(2), 1897–1910.

Claudino, J. G., Capanema, D. d. O., de Souza, T. V., Serrão, J. C., Machado Pereira, A. C., & Nassis, G. P. (2019). Current approaches to the use of artificial intelligence for injury risk assessment and performance prediction in team sports: A systematic review. *Sports Medicine-Open*, *5*, 1–12.

Costa, C. J., Aparicio, M., Aparicio, S., & Aparicio, J. T. (2024). The democratization of artificial intelligence: Theoretical framework. *Applied Sciences*, *14*(18), 8236.

Crevier, D. (1993). *AI: The Tumultuous Search for Artificial Intelligence*. Basic Books.

Degeling, M., Utz, C., Lentzsch, C., Hosseini, H., Schaub, F., & Holz, T. (2018). We value your privacy… now take some cookies: Measuring the GDPR's impact on web privacy. *arXiv preprint arXiv:1808.05096*.

Dindorf, C., Bartaguiz, E., Gassmann, F., & Fröhlich, M. (2022). Conceptual structure and current trends in artificial intelligence, machine learning, and deep learning research in sports: A bibliometric review. *International Journal of Environmental Research and Public Health*, *20*(1), 173.

Du, J., Pifer, N. D., & Gulavani, S. S. (2023). Harnessing the power of machine learning in sport consumer behavior research. In H. K. Leng & J. J. Zhang (Eds.), *Sports Sponsorship and Branding* (pp. 165–183). Routledge.

Duan, Z., Ge, N., & Kong, Y. (2024). The factors affecting aerobics athletes' performance using artificial intelligence neural networks with sports nutrition assistance. *Scientific Reports*, *14*(1), 29639.

Ebbers, F. (2023). Effects of the GDPR in Southeast Asia vs. Europe-A large-scale analysis of IoT devices. In *2023 IEEE 8th International Conference on Software Engineering and Computer Systems (ICSECS)* (pp. 187–194). IEEE.

El-Maghrabi, Y., & Sharif, M. (2022). Game changers or game predictors? Big data analytics in sports for performance enhancement and fan engagement. *Journal of Contemporary Healthcare Analytics*, *6*(6), 19–39.

Elnour, M., Mohammedsherif, H., Fadli, F., Meskin, N., Ahmad, A. M., Rezgui, Y., Petri, I., & Hodorog, A. (2021). Neural network-based predictive control system for energy optimization in sports facilities: A case study. In R. Amor, J. Beetz, & J. McArthur (Eds.), *Proceedings of the 38th CIB W78 Conference on Information and Communication Technologies* (pp. 734–743). CIB.

Ernst, E., Merola, R., & Samaan, D. (2019). Economics of artificial intelligence: Implications for the future of work. *IZA Journal of Labor Policy*, *9*(1), 1–35.

EU. (2024). The AI act explorer. *EU Artificial Intelligence Act*. Retrieved March 4, 2025, from https://artificialintelligenceact.eu/ai-act-explorer/

Fang, C., & Tse, A. W. C. (2022). Quasi-experiment: Postgraduate students' class engagement in various online learning contexts when taking privacy issues to incorporate with artificial intelligence applications. In M. B. Frutos & B. Zhang (Eds.), *Proceedings of the 14th International Conference on Education Technology and Computers* (pp. 356–361). Association for Computing Machinery.

Gabbett, T. J. (2010). The development and application of an injury prediction model for noncontact, soft-tissue injuries in elite collision sport athletes. *The Journal of Strength & Conditioning Research*, *24*(10), 2593–2603.

Galekwa, R. M., Tshimula, J. M., Tajeuna, E. G., & Kyandoghere, K. (2024). A systematic review of machine learning in sports betting: Techniques, challenges, and future directions. *arXiv preprint arXiv:2410.21484*.

Galily, Y. (2018). Artificial intelligence and sports journalism: Is it a sweeping change? *Technology in Society*, *54*, 47–51.

Garvey, C. (2018). A framework for evaluating barriers to the democratization of artificial intelligence. In S. A. McIlraith & K. Q. Weinberger (Eds.), *Proceedings of the Thirty-Second AAAI Conference on Artificial Intelligence* (pp. 123–130). AAAI Press.

Germani, F., Spitale, G., & Biller-Andorno, N. (2024). The dual nature of AI in information dissemination: Ethical considerations. *JMIR AI*, *3*(1), e53505.

Gilbert, C., & Gilbert, M. A. (2024). The convergence of artificial intelligence and privacy: Navigating innovation with ethical considerations. *International Journal*, *3*(9), 9–17.

Glebova, E., Madsen, D. Ø., Mihaľová, P., Géczi, G., Mittelman, A., & Jorgič, B. (2024). Artificial intelligence development and dissemination impact on the sports industry labor market. *Frontiers in Sports and Active Living*, *6*, 1363892.

Goes, F. R., Kempe, M., Meerhoff, L. A., & Lemmink, K. A. (2019). Not every pass can be an assist: a data-driven model to measure pass effectiveness in professional soccer matches. *Big Data*, *7*(1), 57–70.

Goldman, S. (2023, August 30). At the US Open, IBM serves up AI-generated tennis commentary and draw analysis. *VentureBeat*. https://venturebeat.com/ai/ibm-serves-up-ai-generated-tennis-commentary-and-draw-analysis-at-the-us-open/

Gowda, M., Dhekne, A., Shen, S., Choudhury, R. R., Yang, S. X., Yang, L., Golwalkar, S., & Essanian, A. (2018). Iot platform for sports analytics. *GetMobile: Mobile Computing and Communications*, *21*(4), 8–14.

Guéziec, A. (2002). Tracking pitches for broadcast television. *Computer*, *35*(3), 38–43.

Haigh, T. (2023). There was no'first AI winter'. *Communications of the ACM*, *66*(12), 35–39.

Himmelreich, J. (2022). Against "democratizing AI". *AI & Society*, *38*(4), 1333–1346.

Huang, J., Ma, H., Chen, D., Yuan, H., Zhang, J., Li, Z., Han, J., Wu, J., & Yang, J. (2021). Digital nanophotonics: The highway to the integration of subwavelength-scale photonics: Ultra-compact, multi-function nanophotonic design based on computational inverse design. *Nanophotonics*, *10*(3), 1011–1030.

Hughes, J. (2023). *Addressing the Risks of Artificial Intelligence (AI)*. https://www.compassitc.com/blog/addressing-the-risks-of-artificial-intelligence-ai

Hulin, B. T., Gabbett, T. J., Lawson, D. W., Caputi, P., & Sampson, J. A. (2016).

Irfan, M., Aldulaylan, F., & Alqahtani, Y. (2023). Ethics and privacy in Irish higher education: A comprehensive study of Artificial Intelligence (AI) tools implementation at University of Limerick. *Global Social Sciences Review*, *8*(2), 201–210.

Johnson, K. (2020). AI Weekly: AI democratization depends on tech giants. *VentureBeat*. https://venturebeat.com/ai/ai-weekly-ai-democratization-depends-on-tech-giants/

Kakavas, G., Malliaropoulo, N., Mitrotasio, M., Tsaklis, P., & Bikos, G. (2022). Is virtual twins technology the future in sports medicine. *Journal of Physical Medicine Rehabilitation & Disabilities*, *8*(1), 75.

Kaplan, A., & Haenlein, M. (2019). Siri, Siri, in my hand: Who's the fairest in the land? On the interpretations, illustrations, and implications of artificial intelligence. *Business Horizons*, *62*(1), 15–25.

Karami, A., Shemshaki, M., & Ghazanfar, M. (2024). Exploring the ethical implications of AI-powered personalization in digital marketing. *Data Intelligence*, In-Press.

Kaur, D., Ahsan, M., & Gill, G. S. (2025). Digital transformation in the sports industry: driving circular economy in sports equipment and infrastructure. In E. Ozen, A. Singh, S. Taneja, R. Rajaram, & J. Davim (Eds.), *Innovating Sustainability Through Digital Circular Economy* (pp. 65–90). IGI Global Scientific Publishing.

Khan, M. (2023). *Artificial Intelligence: The Next Frontier in Healthcare*. Digital Minds IT. https://www.beawarebettercare.com/2023/07/the-game-changing-effects-of-ai-on.html

Khan, S., Iqbal, M., Osho, O., Singh, K., Derrick, K., Nelson, P., Li, L., Sidnam-Mauch, E., Bannister, N., Caine, K., & Knijnenburg, B. (2024). Teaching middle schoolers about the privacy threats of tracking and pervasive personalization: A classroom intervention using design-based research. In F. F. Mueller & P. Kyburz (Eds.), *Proceedings of the CHI Conference on Human Factors in Computing Systems* (pp. 1–26).

Korzynski, P., Kozminski, A. K., & Baczynska, A. (2023). Navigating leadership challenges with technology: Uncovering the potential of ChatGPT, virtual reality, human capital management systems, robotic process automation, and social media. *International Entrepreneurship Review*, *9*(2), 7–18.

Kovalchik, S. A. (2023). Player tracking data in sports. *Annual Review of Statistics and Its Application*, *10*(1), 677–697.

Layton, R., & Elaluf-Calderwood, S. (2019). A social economic analysis of the impact of GDPR on security and privacy practices. In *2019 12th CMI Conference on Cybersecurity and Privacy (CMI)* (pp. 1–6). IEEE.

Lea, A. S. (2023). *Digitizing Diagnosis: Medicine, Minds, and Machines in Twentieth-Century America*. Johns Hopkins University Press.

LeCun, Y., Bengio, Y., & Hinton, G. (2015). Deep learning. *Nature*, *521*(7553), 436–444.

Lee, K., Han, K., & Ko, J. (2024). Analyzing the impact of the automatic ball-strike system in professional baseball: A case study on KBO league data. *arXiv preprint arXiv:2407.15779*.

Li, R. T., Kling, S. R., Salata, M. J., Cupp, S. A., Sheehan, J., & Voos, J. E. (2016). Wearable performance devices in sports medicine. *Sports Health*, *8*(1), 74–78.

Lingappan, M., Sethu, S., Parasuraman, T., Ramakrishnan, R., Devi, L. S., Reddy, V., & Ranjith, V. (2025). Automating athletic excellence through intelligent process automation and sports analytics. In D. Thangam (Ed.), *Advancements in Intelligent Process Automation* (pp. 179–200). IGI Global.

Liu, Y., Wang, Z., & Hou, W. (2024). Motion measurement data analysis based on AI recognition technology. In *2024 International Conference on Intelligent Algorithms for Computational Intelligence Systems (IACIS)* (pp. 1–5). IEEE.

Madhuri, T., & Sowjanya, P. (2016). Microsoft Azure v/s Amazon AWS cloud services: A comparative study. *International Journal of Innovative Research in Science, Engineering and Technology*, *5*(3), 3904–3907.
Malone, S., Roe, M., Doran, D. A., Gabbett, T. J., & Collins, K. D. (2017). Protection against spikes in workload with aerobic fitness and playing experience: The role of the acute: chronic workload ratio on injury risk in elite Gaelic football. *International Journal of Sports Physiology and Performance*, *12*(3), 393–401.
Marinakis, V., Koutsellis, T., Nikas, A., & Doukas, H. (2021). Ai and data democratisation for intelligent energy management. *Energies*, *14*(14), 4341.
Mazza, B., & Russo, G. (2023). The value of esports football. Towards new models of consumption and participatory experience in Italy. *Contemporary Social Science*, *18*(1), 109–124.
McCarthy, J. (2007). From here to human-level AI. *Artificial Intelligence*, *171*(18), 1174–1182.
McClain, C., Faverio, M., Anderson, M., & Park, E. (2023). *How Americans View Data Privacy*. Pew Research Center.
Musat, C. L., Mereuta, C., Nechita, A., Tutunaru, D., Voipan, A. E., Voipan, D., Mereuta, E., Gurau, T. V., Gurău, G., & Nechita, L. C. (2024). Diagnostic applications of AI in sports: A comprehensive review of injury risk prediction methods. *Diagnostics*, *14*(22), 2516.
Nalbant, K. G. (2021). The importance of artificial intelligence in education: A short review. *Journal of Review in Science and Engineering*, *2021*, 1–15.
Nalbant, K. G., & Aydın, S. (2022). Literature review on the relationship between artificial intelligence technologies with digital sports marketing and sports management. *Indonesian Journal of Sport Management*, *2*(2), 135–143.
Newquist HP. (1994). *The Brain Makers: Genius, Ego, and Greed in the Quest for Machines that Think*. Macmillan.
Nguyen, A., Ngo, H. N., Hong, Y., Dang, B., & Nguyen, B.-P. T. (2023). Ethical principles for artificial intelligence in education. *Education and Information Technologies*, *28*(4), 4221–4241.
Nicolella, D. P., Torres-Ronda, L., Saylor, K. J., & Schelling, X. (2018). Validity and reliability of an accelerometer-based player tracking device. *PloS One*, *13*(2), e0191823.
Nowak, M., Bok, B., Wilczek, A., Oleksy, Ł., & Kamola, M. (2024). Forecasting extremes of football players' performance in matches. *Scientific Reports*, *14*(1), 27319.
Okech, D. (2024). The vital role of AI in driving towards a greener future. *AutoMag.ug*. https://automag.ug/2024/03/27/the-vital-role-of-ai-in-driving-towards-a-greener-future/
Pantelic, O., Jovic, K., & Krstovic, S. (2022). Cookies implementation analysis and the impact on user privacy regarding GDPR and CCPA regulations. *Sustainability*, *14*(9), 5015.
Park, G. (2019). The changing wind of data privacy law: A comparative study of the European Union's general data protection regulation and the 2018 California Consumer Privacy Act. *UC Irvine Law Review*, *10*, 1455.
Patel, R. (2024, February 27). Balancing act: Ethics and privacy in the age of big data analytics. *CIO Influence*. https://cioinfluence.com/it-and-devops/balancing-act-ethics-and-privacy-in-the-age-of-big-data-analytics/
Peranzo, P. (2024). How artificial intelligence is transforming the sports industry. *Imaginovation*. https://imaginovation.net/blog/ai-in-sports-industry/
Peukert, C., Bechtold, S., Batikas, M., & Kretschmer, T. (2022). Regulatory spillovers and data governance: Evidence from the GDPR. *Marketing Science*, *41*(4), 746–768.
Pfleegor, A. G., Katz, M., & Bowers, M. T. (2019). Publish, perish, or salami slice? Authorship ethics in an emerging field. *Journal of Business Ethics*, *156*, 189–208.

Pion-Tonachini, L., Bouchard, K., Martin, H. G., Peisert, S., Holtz, W. B., Aswani, A., Dwivedi, D., Wainwright, H., Pilania, G., & Nachman, B. (2021). Learning from learning machines: A new generation of AI technology to meet the needs of science. *arXiv preprint arXiv:2111.13786*.

Pisaniello, A. (2024). The game changer: How Artificial Intelligence is transforming sports performance and strategy. *Geopolitical, Social Security and Freedom Journal, 7*(1), 75–84.

Pratisto, E. H., Thompson, N., & Potdar, V. (2022). Immersive technologies for tourism: A systematic review. *Information Technology & Tourism, 24*(2), 181–219.

Puce, L., Ceylan, H. İ., Trompetto, C., Cotellessa, F., Schenone, C., Marinelli, L., Zmijewski, P., Bragazzi, N., & Mori, L. (2024). Optimizing athletic performance through advanced nutrition strategies: can AI and digital platforms have a role in ultraendurance sports? *Biology of Sport, 41*(4), 305–313.

Rafalski, K. (2024). AI in sports: Transforming the game for players and fans. *Netguru*. https://www.netguru.com/blog/ai-in-sports

Rahmani, M., Majedi, N., Hemmatinejad, M., & Jamshidi, A. (2024a). Application of Artificial Intelligence in the Sports Industry: A Review Article. *AI and Tech in Behavioral and Social Sciences, 2*(2), 20–27.

Rahmani, N., Naderi Nasab, M., Taheri, M., & Biniaz, S. A. (2024b). The future of sports industry: AI and economic transformations. *AI and Tech in Behavioral and Social Sciences, 2*(4), 19–29.

Ramkumar, P. N., Luu, B. C., Haeberle, H. S., Karnuta, J. M., Nwachukwu, B. U., & Williams, R. J. (2022). Sports medicine and artificial intelligence: A primer. *The American Journal of Sports Medicine, 50*(4), 1166–1174.

Rangasamy, K., As'ari, M. A., Rahmad, N. A., Ghazali, N. F., & Ismail, S. (2020). Deep learning in sport video analysis: A review. *TELKOMNIKA (Telecommunication Computing Electronics and Control), 18*(4), 1926–1933.

Ráthonyi, G., Müller, A., & Rathonyi-Odor, K. (2018). How digital technologies are changing sport? *Applied Studies in Agribusiness and Commerce, 12*(3–4), 89–96.

Resnik, D. B. (2015). What is ethics in research & why is it important. *National Institute of Environmental Health Sciences*. https://www.niehs.nih.gov/research/resources/bioethics/whatis

Rossi, A., Pappalardo, L., Cintia, P., Iaia, F. M., Fernández, J., & Medina, D. (2018). Effective injury forecasting in soccer with GPS training data and machine learning. *PloS One, 13*(7), e0201264.

Ruddy, J. D., Shield, A. J., Maniar, N., Williams, M. D., Duhig, S. J., Timmins, R. G., Hickey, J., Bourne, M. N., & Opar, D. A. (2018). Predictive modeling of hamstring strain injuries in elite Australian footballers. *Medicine & Science in Sports & Exercise, 50*(5), 906–914.

Russell, S. J., & Norvig, P. (Eds.) (1995). *Artificial intelligence: A Modern Approach*. Upper Saddle River, NJ: Prentice Hall.

Santos-Gago, J. M., Ramos-Merino, M., Vallarades-Rodriguez, S., Álvarez-Sabucedo, L. M., Fernández-Iglesias, M. J., & García-Soidán, J. L. (2019). Innovative use of wrist-worn wearable devices in the sports domain: A systematic review. *Electronics, 8*(11), 1257.

Saraswat, M., & Tripathi, R. C. (2020). Cloud computing: Comparison and analysis of cloud service providers—AWS, Microsoft, and Google. In R. K. Dwivedi (Ed.), *2020 9th International Conference System Modeling and Advancement in Research Trends (SMART)* (pp. 281–285). IEEE.

Sarioguz, O. & Miser, E. (2024). Assessing the role of artificial intelligence in enhancing customer personalization: A study of ethical and privacy implications in digital marketing. *International Journal of Science and Research Archive, 13*(2), 812–825.

Sarkar, M., Puja, A. R., & Chowdhury, F. R. (2024). Optimizing marketing strategies with RFM method and K-Means clustering-based AI customer segmentation analysis. *Journal of Business and Management Studies*, *6*(2), 54–60.

Shah, N. (2024). Artificial intelligence and leadership: how artificial intelligence is changing the leadership role. *Remittances Review*, *9*(1), 2750–2764.

Shortliffe, E. H., & Buchanan, B. G. (1975). A model of inexact reasoning in medicine. *Mathematical Biosciences*, *23*(3–4), 351–379.

Shrestha, Y. R., Ben-Menahem, S. M., & von Krogh, G. (2019). Organizational decision-making structures in the age of Artificial Intelligence. *California Management Review*, *61*(4), 66–83.

Swant, M. (2024, August 27). How IBM and the US open are using Watsonx to create more AI-generated tennis content. *Digiday*. https://digiday.com/media/how-ibm-and-the-us-open-are-using-watson-create-more-ai-generated-tennis-content/

Thilagavathi, K. (2024). AI-driven customer segmentation: A targeting and retargeting strategies in digital marketing. *Journal of Dynamics and Control*, *8*(8), 400–415.

Trail, G., Kim, A., Bang, H., & Braunstein-Minkove, J. R. (2024). Ethics in quantitative sport management research: The impact of AI. *International Journal of Sports Marketing and Sponsorship*, *25*(5), 1147–1162.

Tseng, B.-Y., Chiang, H.-T., Chen, J.-L., & Hsieh, H.-C. (2024). Pitching-motion: Pose-based pitch trajectory overlay system. In *2024 26th International Conference on Advanced Communications Technology (ICACT)* (pp. 116–121). IEEE.

Tussyadiah, I. P., & Pesonen, J. (2016). Impacts of peer-to-peer accommodation use on travel patterns. *Journal of Travel Research*, *55*(8), 1022–1040.

Tyler, N. S., Mosquera-Lopez, C., Young, G. M., El Youssef, J., Castle, J. R., & Jacobs, P. G. (2022). Quantifying the impact of physical activity on future glucose trends using machine learning. *Iscience*, *25*(3), 1–19.

Ucuz, D. (2020, June). Comparison of the IoT platform vendors, Microsoft Azure, Amazon Web Services, and Google Cloud, from users' perspectives. In M. Karabatak, F. Ertam, & A. Varol (Eds.), *2020 8th International Symposium on Digital Forensics and Security (ISDFS)* (pp. 1–4). IEEE.

Van Eetvelde, H., Mendonça, L. D., Ley, C., Seil, R., & Tischer, T. (2021). Machine learning methods in sport injury prediction and prevention: A systematic review. *Journal of Experimental Orthopaedics*, *8*, 1–15.

Wang, Y. (2021). When artificial intelligence meets educational leaders' data-informed decision-making: A cautionary tale. *Studies in Educational Evaluation*, *69*, 100872.

White House. (2023). Blueprint for an AI bill of rights | OSTP. *The White House*. https://bidenwhitehouse.archives.gov/ostp/ai-bill-of-rights/

Xie, H., & Xu, L. (2024). Intelligent recognition of Athlete's erroneous movements in sports training under artificial intelligence technology. In *2024 Third International Conference on Distributed Computing and Electrical Circuits and Electronics (ICDCECE)* (pp. 1–6). IEEE.

Xu, T., & Baghaei, S. (2025). Reshaping the future of sports with artificial intelligence: Challenges and opportunities in performance enhancement, fan engagement, and strategic decision-making. *Engineering Applications of Artificial Intelligence*, *142*, 109912.

Zaman, B. U. (2024). *Leveraging big data and AI for personalized learning opportunities, challenges, and ethical considerations, challenges, and ethical considerations*. Preprints.org doi: 10.20944/preprints202407.1211.v1

Zaman, R., Cuzzocrea, A., & Hassani, M. (2019). An innovative online process mining framework for supporting incremental GDPR compliance of business processes. In *2019 IEEE International Conference on Big Data (Big Data)* (pp. 2982–2991). IEEE.

6

VIRTUAL REALITY AND AUGMENTED REALITY IN SPORT

Chad Goebert and John David Johnson

Introduction

Augmented Reality (AR) and Virtual Reality (VR) are continually evolving and expanding technological fields. AR is a computer technology that superimposes digital images or data on pre-existing environments (Abeza et al., 2024). VR is the computer-generated simulation of a three-dimensional image or environment with which a person can interact in a seemingly real or physical way using special electronic equipment, such as a headset with a screen inside (Abeza et al., 2024). Collectively, they constitute a significant subset within the broader category referred to as immersive technologies. Immersive technologies are technologies that blur the line between the virtual and physical worlds to the point that users feel as though they are physically present in their digitally created surroundings (Suh & Prophet, 2018). According to data from research conducted by Statista in cooperation with ARrtillery Intelligence, a spatial computing research and insights company, the global market for immersive technologies was valued at 25.7 billion dollars in 2024, with forecasts suggesting a market value of 41.8 billion dollars by 2028 (Alsop, 2025). Credence Research (2024) projects similar growth for the use of AR and VR specifically in the sport industry. Industries such as gaming, retail, healthcare, military, education, and sport are increasing their use of AR and VR for training, diagnostics, marketing, and enhanced learning (Ahir et al., 2020; Desai et al., 2024). Improvements in hardware, software, and computing power have allowed AR and VR to become more accessible, efficient, and user-friendly (Al-Ansi et al., 2023).

The widespread adoption of AR and VR technologies has also been accelerated by substantial investments from major corporations, including Apple, Microsoft, Facebook, Google, Amazon, and Disney (Marr, 2021). Along with the previously mentioned technology and business giants, major sports organizations, and leagues such as the National Football League (NFL), National Hockey League (NHL), Major League Soccer (MLS), Fédération Internationale

DOI: 10.4324/9781003588405-8

de Football Association (FIFA), Major League Baseball (MLB), National Basketball Association (NBA), and many others have also begun incorporating AR and VR technologies to enhance fan engagement and deliver immersive viewing experiences (Potwarka et al., 2022; Young, 2020). The incorporation of these technologies by leading sport organizations underscores the need to understand how they are being used and how they will potentially be used to enhance the sport experience for spectators and athletes alike. AR and VR are already being used in sport. The NBA offers virtual courtside seats, and the NFL has used AR effects in broadcasts aimed at younger fans, like the Nickelodeon games. On the training side, baseball players use WIN Reality to practice hitting against simulated pitching, and clubs like Manchester City use Rezzil's VR system to help players improve decision-making and awareness. These are not just future ideas; they are already part of how teams train and how fans consume sport.

Background

VR was first introduced in studies dating back to at least the 1960s (Handa et al., 2012). At that time, the technology was cumbersome and inaccessible to most. However, improvements in display technology, motion tracking, and VR software have led to the creation of immersive environments that were once unattainable. These improvements allow organizations and companies in the sport industry to build training tools and simulations (Rezzil, Win Reality, Xtadium NBA VR, etc.) that mimic real-world scenarios, giving athletes and fans better opportunities to develop skills, consume sport content, and improve performance (Clavijo et al., 2024; Fortes et al., 2021; Potwarka et al., 2022). AR, while different from VR, has followed a similar path of technological progress. AR enhances physical reality by overlaying digital information onto real-world environments (Berryman, 2012). The key difference between VR and AR is that VR immerses the user in an entirely digitally created environment while AR integrates digitally created elements into the real-world environment.

As previously noted, the increased use of AR and VR technologies is in large part driven by the increase in accessibility and affordability of AR and VR technologies. VR is commonly delivered via a head-mounted display (HMD) (LaValle, 2023). HMDs are eyeglasses, helmets, or headgear that use projection technology to display the digital environment to the user (Renganayagalu et al., 2021). AR, however, is delivered in multiple modalities including HMDs, web AR, projector-based systems, broadcast AR, and smartphone-based AR. (Syed et al., 2022). AR and VR reflect decades of creative inspiration and innovation. While the origins of the two technologies differ, both have grown from niche technologies into practical tools utilized across many industries (Ahir et al., 2020; Desai et al., 2024).

Topical Exploration and Key Concepts

Immersive Training Environments

VR is a powerful tool for creating training environments that mimic real-world scenarios. In football, quarterbacks have been using VR for skill training for years (Huang et al., 2015). Most recently, NFL quarterback Jayden Daniels has openly discussed the importance of VR in his skill training regimen (Hayden, 2024). Skill training through repetitions is one of the most recognized strengths of VR (Richlan et al., 2023; Xie et al., 2021). In sport, VR provides athletes with realistic environments to experience repetitive practice, allowing them to strengthen motor skills, cognitive skills, and psychological skills in a controlled environment (Richlan et al., 2023). Research has shown that VR can improve training across multiple sports. Gray (2017) found that VR training for batting practice boosted performance beyond real-world training alone for baseball players. High school players trained in VR sessions where pitch speed, crossing height, and lateral location were adjusted to match their skill levels. The result was higher on-base percentages and more players advancing to higher competition levels. Similarly, athletes have shown improvements in reaction time in sports ranging from karate to table tennis (Michalski et al., 2019; Petri et al., 2019).

One of the top VR training programs that is commercially available is WIN Reality. WIN Reality lets hitters practice against realistic, game-speed pitching from professional, college, or custom pitcher profiles. WIN Reality offers individualized training that mimics real game conditions and allows batters to experience far more reps in much less time than traditional batting practice, resulting in increased pitch evaluation accuracy (Clavijo et al., 2024). Occlusion exercises help hitters improve pitch recognition and strike zone discipline. WIN Reality data shows consistent improvements in batting averages by about 35 points, On-base Plus Slugging (OPS) increasing by 50 points, and swing ratios from ~2.55 to ~2.90 (Goldman, 2020).

VR training also helps to improve the cognitive and perceptual skills that athletes rely on. Fortes et al. (2021) found that youth soccer players who used VR to view offensive plays from a first-person perspective improved decision-making and visual search behaviors more than players using traditional video training. Rezzil's VR soccer platform can help to improve passing accuracy, composure, reaction time, and adaptability in players, both novice and professional (Wood et al., 2021). The Rezzil Index accurately measures the skill levels of players and adjusts their VR training accordingly. Rezzil was also shown to help players improve their sport-specific perceptual-cognitive skills while allowing them to avoid the physical challenges that accompany traditional training (Gürbüz & Taş, 2025).

VR has even been shown to enhance psychological resilience and motivation. McClure and Schofield (2019) showed that VR-enhanced exercise environments increased focus and workout duration, even when participants felt more physically exerted. Richlan et al. (2023) noted that VR could elicit real effects in sports performance enhancement. They found that VR training groups consistently outperformed control groups that did not utilize VR in their training. Overall, these findings show VR training environments can have real-world impacts on sport performance. By simulating realistic scenarios, VR helps athletes enhance motor, cognitive, and psychological skills better than traditional methods, making VR a valuable tool in modern sports training.

AR and Live Sports

Broadcast AR has gained popularity in live sports broadcasts, with organizations like the NBA, NFL, and English Premier League using AR to enhance viewer engagement through data overlays and real-time analytics (Goebert, 2020). Real-time data and graphics are integrated into traditional sport broadcasts via digitally created elements. AR broadcast technology can provide viewers with more informative and engaging experiences by providing both hedonic and utilitarian information to the viewer in real time (Goebert et al., 2022). AR serves as both an analytical tool for in-depth understanding and an entertainment device for creating more immersive broadcasts.

Perhaps the earliest and most well-known example of AR use in sports broadcasting is the virtual first-down line. The line was first introduced in American football broadcasts in the late 1990s. The simple yet effective graphic allowed viewers to immediately comprehend the progress of a play without relying entirely on verbal explanations from commentators. It has become not only a helpful part of football broadcasts but an expected one (Stern, 2024). From that seemingly simple starting point, AR technology has expanded to various sport broadcasts across the world. Common uses of AR graphics in sports broadcasting include, but are not limited to:

NFL and College Football: Virtual first-down lines, route diagrams, and player statistics displayed during broadcasts on networks such as ESPN and CBS.

Tennis and Cricket: Hawk-Eye technology provides real-time ball tracking and accurate line-calling systems.

NBA Games: Second Spectrum's *CourtVision* technology, which offers both analytical "coach-mode" graphics and entertainment-driven "mascot-mode" overlays.

NHL Games: Enhanced puck-tracking graphics and player identification features designed to simplify the fast-paced nature of hockey broadcasts.

Nickelodeon NFL Broadcasts: Since 2021, these broadcasts have incorporated playful AR visuals like slime cannons, animated scoreboards, and cartoon characters to appeal to younger audiences.

Disney & ESPN's Toy Story Funday Football: An innovative broadcast that recreated an NFL game within the *Toy Story* universe, featuring animated player avatars and commentary from well-known characters.

Soccer (MLS, Premier League, etc.): Real-time player heat maps, passing trajectories, and possession graphics generated by systems including Hawk-Eye, Vizrt, Premier League Data Zone, and others, enhancing both tactical comprehension and viewer engagement.

On the technology side, companies like Hawk-Eye Innovations are continually improving their AR technology. Hawk-Eye was first made to assist with accurate line calls in tennis and cricket. It's now being used for other sports like basketball and soccer, as well. It creates graphics that show heat maps, defensive positioning, and predictions about what might happen next. These graphics help broadcasters tell better stories and give viewers a clearer understanding of the game (Liu et al., 2022).

Not all AR graphics are popular with everyone. A study by Goebert and Greenhalgh (2020) looked at three different ways NBA games were shown on TV. The first type was a regular broadcast without any AR graphics. The second type was called "coach mode," which used AR graphics to help people understand game strategies better. The third type was called "mascot mode," which used flashy, cartoon-like graphics that looked like something from a video game. The results showed that most people liked the "coach-mode" broadcast better than the "mascot-mode." They were also more likely to watch the "coach-mode" broadcast again and recommend it to others. The researchers found that in that instance, AR graphics that help people understand the game were preferred over graphics made just for entertainment.

Younger audiences are consuming sport broadcasts differently than older generations (Gilal et al., 2021; Meneses et al., 2022). AR broadcasts aimed at younger audiences have demonstrated the potential of the technology to resonate with younger audiences. It is important to note that younger audiences are generally more receptive to entertainment-driven AR integrations in sport broadcasts (Goebert et al., 2022). The generational difference in AR acceptance reflects broader trends in digital viewership habits. Younger viewers who potentially grew up using platforms like YouTube, TikTok, and Twitch seem to prefer exciting visuals and interactive content. Older viewers, who are used to more traditional TV broadcasts, do not appear to be as interested in enhanced graphics. Increasingly, broadcasters who want to keep younger audiences watching their content are turning to AR to attempt to make sports broadcasts more fun and engaging. Nickelodeon's NFL broadcasts have successfully attracted children and teens by incorporating cartoonish AR graphics such as slime cannons, playful scoreboards, and animated characters. The broadcasts began in 2021, with viewership numbers ranging from 900,000 to 2.06 million unique viewers per broadcast (Cohen, 2023). The main focus of the broadcasts is to attract younger viewers and make watching football more fun and appealing for younger audiences.

Overall, AR's ability to integrate real-time data and analytical graphics into sports broadcasts represents a significant advancement in how audiences can choose to experience broadcasts. While traditional viewers may prefer analytical AR features aimed at enhancing comprehension, younger audiences appear more receptive to fun, entertaining graphics. As AR technology continues to evolve, it is likely to play an increasingly important role in sports broadcasting, offering a customizable range of experiences designed to meet the varied preferences of sport audiences.

Fan Engagement and VR/AR

Fan engagement has always been a central concern in sport, but emerging technologies such as AR and VR have redefined what it means to connect with a team, player, or sport. VR and AR are creating immersive, personalized, and interactive fan experiences that could potentially deepen emotional connections and expand access to the sport product. One of the most prevalent uses of AR in sport fan engagement is through advertising and promotional campaigns. AR activations not only serve to market products but also drive interaction with fans in new and exciting ways. For example, the College Football Playoff (CFP) and Nike collaborated on an AR scavenger hunt in New Orleans. Fans used the CFP app to locate and scan murals throughout the city, triggering team-specific virtual hype videos and unlocking the chance to purchase limited-edition Nike shoes (College Football Playoff, 2020). The campaign not only leveraged AR for ecommerce but also turned the host city itself into an interactive event space, drawing fans into both the physical and digital environments.

Similarly, the Washington Capitals introduced an AR-enhanced cereal box for "Ovi O's," featuring an interactive hockey game activated via Snapchat. Fans could play as Alex Ovechkin, shooting puck-shaped cereal at virtual targets. The Capitals reported that 20% of consumers who purchased the cereal engaged with the AR experience, demonstrating how these technologies can drive product interaction and brand visibility (Santana, 2019). AR has also been used to enhance the in-venue experience. The Dallas Cowboys' "Pose with the Pros" activation allowed fans at AT&T Stadium to take virtual selfies with players via interactive columns throughout the venue. The campaign generated over 50 million social media impressions in its opening weekend alone (Draper, 2019). Such activations turn fans into content creators, amplifying their reach and emotional investment.

VR can offer even deeper levels of immersion by transporting fans into virtual stadiums. The NBA's Virtual Fan Experience during the COVID-19 pandemic is a notable example, allowing fans to watch games in VR and feel "present" in the arena despite the lack of physical crowds. These technologies have proven particularly effective at attracting new audiences and engaging fans who may not otherwise attend games in person (Chung et al., 2024). Research shows that AR and VR can significantly increase emotional engagement, brand loyalty, and satisfaction among sports fans (Chen & Wang, 2018). Overall, VR and AR technologies have moved from being technological novelties to integral components of the fan experience. They offer new ways to interact with content, create lasting impressions, and generate user-driven media. Whether through in-venue activations, gamified mobile apps, immersive broadcasts, or virtual event attendance, VR and AR are reshaping the boundaries of what it means to be a fan in the digital age.

Data-Driven Feedback and Performance Enhancement

Data-driven feedback using VR and AR provides athletes and coaches with insights into athletic performance, skill technique, and rehabilitation progress. These technologies integrate advanced sensors and software capable of tracking biomechanical parameters such as movement patterns, reaction times, and muscular activation, converting the metrics into actionable, real-time data (Soltanabadi et al., 2023). Immediate biofeedback within immersive VR and AR environments facilitates targeted corrections during simulated sport-specific gameplay or controlled rehabilitation tasks, ultimately enhancing athletes' precision and efficiency. VR and AR-based platforms frequently incorporate analytics dashboards that interpret performance data and suggest individualized training programs (Liu et al., 2024). For example, tennis-specific VR applications might employ wearable sensors to assess racquet angles, footwork precision, and upper-body rotation, guiding athletes toward sharper techniques. Similarly, in injury rehabilitation, real-time 3D motion capture can detect imbalances or compensatory movements, visualizing them clearly through HMDs. Such structured data processing supports tracking improvements and informs the effective structuring of subsequent training or rehab sessions (Liu et al., 2024).

A notable strength of VR and AR lies in the targeted development of complex motor skills. Studies employing VR-based test batteries demonstrate reliable tracking of dynamic parameters such as vertical jump height, reactive agility, and multi-segment coordination, showing strong correlations with real-world performance metrics (Pastel et al., 2025). VR environments can dynamically adjust task difficulty based on athlete performance levels, escalating challenges progressively to sustain skill advancement, refine decision-making abilities, and reinforce correct movement patterns (Pastel et al., 2025;

Richlan et al., 2023). Ultimately, the real-time, data-driven capabilities of VR and AR systems empower athletes to optimize both cognitive and physical skill dimensions. Studies highlight measurable improvements in precision, decision-making, and stress management derived from structured virtual simulations (Richlan et al., 2023).

Practice in Action

To gain deeper insight into the application of VR and AR technologies in sport, an interview was conducted with Andy Etches, Co-Founder and Sports Director at Rezzil. Etches has been a pioneer in the field, beginning his work with immersive technologies during his time with Manchester City and later founding Rezzil to advance VR-based athlete development.

Interviewer: How would you characterize the pace of innovation in the VR and AR sports technology space?

Andy Etches: It's a whirlwind, to say the least. There's always something new happening, whether that's a breakthrough in hardware, improvements in software, or new ideas about how to apply the technology in sport. It's exciting but also challenging because the pace is relentless. If you aren't ready to adapt quickly, you can easily fall behind.

Interviewer: How did you first become involved with VR and AR applications in sport?

Andy Etches: My interest really started when I was at Manchester City back in 2013–2014. We had early access to Google Glass and worked with Google to build a small project that eventually made it onto a live broadcast during a match between Manchester City and Liverpool at Yankee Stadium. It was groundbreaking at the time. Even before that, I had a dream of using vision-driven technologies in sport, but that project was one of the first times it really became practical. Later, I started my own business focused on making data more accessible, and that eventually led me toward working more deeply with VR. Through that journey, I connected with people who had developed VR ball-striking experiences, and that collaboration ultimately became the foundation for Rezzil.

Interviewer: What do you see as the most promising current applications of VR and AR in sport?

Andy Etches: In VR, the ability to recreate realistic environments—like feeling the pressure of a packed stadium or seeing an opponent bearing down on you at game speed—is hugely promising. It helps elite athletes visualize and rehearse scenarios before they happen in competition. We've also seen VR help coaches build empathy with players. For example, Gareth Southgate, former England manager, used VR to review a player's decision during a match and said it completely changed his perspective, helping him understand the player's choice in real time. AR has massive fitness potential, too. We have projects in the works that we're holding back until the hardware is ready, but when it is, you'll see some truly wild applications in both training and fan engagement.

Interviewer: How has the technology evolved over the past few years?

Andy Etches: VR has evolved in important ways; devices are getting lighter, more powerful, and less reliant on external hardware. The move toward wireless headsets and dedicated, application-specific devices is making it easier to integrate into sports settings. AR has evolved more slowly and has been focused mostly on consumer experiences so far. Still, even with current limitations, the potential is clear.

Interviewer: Let's shift to Rezzil specifically. What was the original purpose of the platform, and how has it evolved?

Andy Etches: Originally, Rezzil was designed as a talent identification tool to help clubs and coaches scout young players. VR allowed us to create repeatable, standardized scenarios for evaluation, something very difficult to achieve in real-world settings. Over time, with feedback from elite coaches and players, the platform expanded into injury rehabilitation, mental skills development, and cognitive training. Today, we have multiple tailored products, everything from tools for elite rehab to modules that help players improve decision-making and anticipation skills under game-like pressure.

Interviewer: What would you say are the biggest challenges or barriers to wider adoption of VR and AR technologies in sport?

Andy Etches: The weight of the headsets is still a major issue, especially for physically dynamic sports. Until headsets are lighter and better balanced, there will be limits on how athletic the VR training environments can get. There's also cultural resistance; coaches and athletes are understandably cautious about adopting new tools without long-term validation. Financially, cost is a barrier for amateur clubs, but elite organizations are willing to invest once they see results. Academic validation also takes time, which can slow broader acceptance.

Interviewer: Looking ahead, what trends or advancements do you anticipate shaping the next five to ten years in VR and AR for sport?

Andy Etches: Devices will continue to get smaller, lighter, and more specialized. I also expect to see a lot more success stories in mainstream sports—stories like NFL quarterbacks using VR for cognitive training, which will drive broader acceptance. AI will play a much bigger role, too, with smarter systems that adapt to athlete performance in real-time. On the AR side, we'll see more halfway technologies, things like camera-driven wearables or glasses that display simple overlays, until full AR glasses become truly viable. It's going to be incremental, but the potential is huge.

Interviewer: For organizations or researchers just beginning to explore immersive technologies, what advice would you offer?

Andy Etches: Start small. Pick up an off-the-shelf headset like the Meta Quest and try out some of the excellent tools already available. There are VR fitness experiences, cognitive training programs, and plenty of performance-focused apps ready to use. Let your team experiment with what's out there. You might be surprised by what's already possible. Most likely, the tool or capability you're imagining already exists or is close to being achievable.

Trends and Current Issues

Advances in Hardware, Software, and AI

VR & AR are technologies that are steadily advancing in quality, thanks in large part to improvements in hardware, software, and artificial intelligence (AI). Today's VR headsets feature cutting-edge display technology, like Organic Light-Emitting Diode (OLED) screens that deliver stunning, lifelike visuals. The latest models also feature wider fields of view, which help make digital environments feel more immersive. Improved refresh rates help to reduce lag and motion sickness, providing a smoother experience. Devices like Meta Quest 3 and Apple Vision Pro are becoming smaller, lighter, and free of cables, which makes them easier to use for longer periods.

It's not just hardware improvements that have advanced; AR and VR software developers are continually innovating and pushing realism to new heights. Advanced software now provides more detailed lighting, shadows, and physics interactions, making virtual graphics more closely mimic real-world settings (Gonçalves et al., 2022). Techniques like photogrammetry turn real-world photos into high-quality 3D models, streamlining content creation without sacrificing detail. Gesture and motion controls have also evolved, allowing for natural, intuitive interactions without the need for bulky controllers (Schäfer et al., 2022). Multi-user connectivity has taken off as well, letting people connect and collaborate in shared virtual spaces. With improved software for smartphones, computers, and HMDs, more people have access to high-quality AR and VR than ever before.

AI plays a pivotal role in the improvement of AR and VR technology. It powers more realistic virtual environments by generating detailed content efficiently, making production faster and more cost-effective (Devagiri et al., 2022). AI is especially useful in sport uses of AR and VR to analyze large amounts of data and process that data into real-time guidance during complex tasks (Cossich et al., 2023). AI's ability to recognize and track objects also allows virtual elements to blend seamlessly into the real world, making AR and VR feel less like novelty gimmicks and more like practical tools. The integration of AI with AR and VR is making experiences more intuitive and personalized. From creating adaptive environments that adjust content in real-time to providing more dynamic user interactions, AI is adding a layer of intelligence that makes these technologies feel much more immersive (Cossich et al., 2023). Thanks to the advances in hardware, software, and AI, AR and VR are becoming an integral part of athlete training and sport fan experiences (Le Noury et al., 2022).

Barriers to Widespread Adoption

There are multiple barriers and potential issues regarding the widespread adoption and use of AR and VR in Sports. Although AR and VR provide a medium in which to enhance the sport training and spectator experience, there are still issues that make it challenging for consumers to adopt and regularly use AR and VR in sport.

Cost

AR and VR systems can be very expensive. Buying the equipment, paying for software, keeping everything updated, and hiring people who know how to use the technology can be costly. The good news is that the costs to implement and use these technologies have been steadily decreasing (Al-Ansi et al., 2023).

Technology Accessibility Issues

Effective use of AR and VR in sports depends on more than just access to devices. Many AR and VR systems require high-speed, stable internet connections, significant computing power, and a robust IT infrastructure (Creed et al., 2024). In addition, coaches, athletes, and staff often need specialized training to understand and operate the technology effectively. Learning how to operate both hardware and software systems can present a steep learning curve. Without the necessary technical support and digital literacy, the adoption and integration of AR and VR tools can be challenging (Chang et al., 2023).

Finding the Best Fit

To get real value from AR and VR, sports organizations must focus on purposeful use, not just adopting tech for the sake of novelty. Success depends on cost-effective solutions, reliable access to technology, and clear sport-specific use cases (Le Noury et al., 2022; Lewis, 2024).

Summary

AR and VR, as part of the broader category of immersive technologies, are impacting the sport industry. Their uses in sport range from athlete training and injury rehabilitation to broadcast enhancements and fan engagement. The chapter outlines the historical evolution of AR and VR, details improvements in hardware, software, and AI, and highlights their growing integration across major leagues like the NFL, NBA, and FIFA.

Despite the potential of the technologies, widespread adoption of AR and VR still face significant challenges. Costs and the need for specialized technical knowledge can create barriers to use. Skepticism about the practical benefits of AR and VR in sport further underscores the need for research-backed industry uses of the technology. Adoption should not be driven by innovation alone but by strategic use cases that make sense within specific sport contexts. By balancing innovation with practicality, the sport industry can fully leverage AR and VR to improve both athlete performance outcomes and the spectator experience.

Key Resources

- **Genius Sports** (https://www.geniussports.com/engage/augmentation/)
- **Rezzil** (https://rezzil.com/)
- **Win Reality** (https://winreality.com/)
- **Balti Virtual** (https://www.baltivirtual.com/)
- **The Famous Group** (https://www.thefamousgroup.com/)
- **@ARVRinSport** (https://x.com/ARVRinSport)

References

Abeza, G., Wanless, E., & Mamo, Y. (2024). Emerging developments in sport management. In Pedersen, P. & Thibault, L. (Eds.), *Contemporary Sport Management* (pp. 348–350. HK.

Ahir, K., Govani, K., Gajera, R., & Shah, M. (2020). Application on virtual reality for enhanced education learning, military training and sports. *Augmented Human Research*, *5*, 1–9.

Al-Ansi, A. M., Jaboob, M., Garad, A., & Al-Ansi, A. (2023). Analyzing augmented reality (AR) and virtual reality (VR) recent development in education. *Social Sciences & Humanities Open*, *8*(1), 100532.

Alsop, T. (2025). Extended reality (XR) market revenue worldwide from 2023 to 2028. https://www.statista.com/statistics/591181/global-augmented-virtual-reality-market-size/

Berryman, D. R. (2012). Augmented reality: A review. *Medical Reference Services Quarterly*, *31*, 212–218.

Chang, C. Y., Kuo, H. C., & Du, Z. (2023). The role of digital literacy in augmented, virtual, and mixed reality in popular science education: A review study and an educational framework development. *Virtual Reality*, *27*(3), 2461–2479.

Chen, L., & Wang, H. (2018). The impact of virtual reality on spectator experience at live sporting events. *Journal of Sport Management*, *32*(5), 419–434.

Chung, K. S., Goebert, C., & Johnson, J. D. (2024). Virtual reality and higher education sporting events: Social anxiety perception as an outcome of VR simulation. *Behavioral Sciences*, *14*(8), 695.

Clavijo, F. A. R., Romeas, T., & Trempe, M. (2024). Effect of batting practice in virtual reality on Pitch recognition in highly trained baseball players. *Journal of Exercise, Movement, and Sport (SCAPPS Refereed Abstracts Repository)*, *55*(1). https://www.scapps.org/jems/index.php/1/en/issue/view/17

Cohen, A. (2023, August 1). NFL returns to Nickelodeon with alternate Super Bowl telecast. *Front Office Sports*. https://frontofficesports.com/nfl-returns-to-nickelodeon-with-first-alternate-super-bowl-telecast/

College Football Playoff. (2020). College Football Playoff teams up with Nike for augmented reality experience at 2020 national championship game. *College Football Playoff*, January 10. https://collegefootballplayoff.com/news/2020/1/10/cfp-nike-ar-activation.aspx

Cossich, V. R., Carlgren, D., Holash, R. J., & Katz, L. (2023). Technological breakthroughs in sport: Current practice and future potential of artificial intelligence, virtual reality, augmented reality, and modern data visualization in performance analysis. *Applied Sciences*, *13*(23), 12965.

Credence Research. (2024, June 4). Sports AR and VR market size, growth & forecast to 2032. https://www.credenceresearch.com/report/sports-augmented-reality-ar-and-virtual-reality-vr-market

Creed, C., Al-Kalbani, M., Theil, A., Sarcar, S., & Williams, I. (2024). Inclusive AR/VR: Accessibility barriers for immersive technologies. *Universal Access in the Information Society*, *23*(1), 59–73.

Desai, K., Kumar, M. D., Khairi, M. O., Govindarajo, N. S., & Sharma, V. (2024). Virtual reality and augmented reality for industrial and business intelligence. In *Service Innovations in Tourism: Metaverse, Immersive Technologies, and Digital Twin* (pp. 64–85). IGI Global Scientific Publishing.

Devagiri, J. S., Paheding, S., Niyaz, Q., Yang, X., & Smith, S. (2022). Augmented reality and artificial intelligence in industry: Trends, tools, and future challenges. *Expert Systems with Applications*, *207*, 118002.

Draper, L. (2019, September 12). Fans excited to "pose with the pros". *Dallas Cowboys*. https://www.dallascowboys.com/news/fans-excited-to-pose-with-the-pros

Fortes, L. S., Almeida, S. S., Praça, G. M., Nascimento-Júnior, J. R., Lima-Junior, D., Barbosa, B. T., & Ferreira, M. E. (2021). Virtual reality promotes greater improvements than video-stimulation screen on perceptual-cognitive skills in young soccer athletes. *Human Movement Science*, *79*, 102856.

Gilal, F. G., Gilal, N. G., Tariq, B., Gilal, R. G., Gilal, R. G., Gong, Z., & Channa, N. A. (2021). Intergenerational differences in fans' motivation to watch the T20 World Cup: A generation cohort theory perspective. *International Journal of Sports Marketing and Sponsorship*, *22*(2), 197–219.

Goebert, C. (2020). Augmented reality in sport marketing: Uses and directions. *Sports Innovation Journal*, *1*, 134–151.

Goebert, C., Greenhalgh, G., & Dwyer, B. (2022). A whole new ball game: Fan perceptions of augmented reality enhanced sport broadcasts. *Computers in Human Behavior*, *137*, 107388.
Goebert, C., & Greenhalgh, G. P. (2020). A new reality: Fan perceptions of augmented reality readiness in sport marketing. *Computers in Human Behavior*, *106*, 106231.
Goldman, M. (2020). Elite baseball training in virtual reality. https://winreality/?.com/blog/elite-baseball-training-virtual-reality/??srsltid=AfmBOopLM0WiY6jLmz2xezJNfoVqicZSayfhdzfZlbmOXt7f5uMpr-Nq
Gonçalves, G., Coelho, H., Monteiro, P., Melo, M., & Bessa, M. (2022). Systematic review of comparative studies of the impact of realism in immersive virtual experiences. *ACM Computing Surveys*, *55*(6), 1–36.
Gray, R. (2017). Transfer of training from virtual to real baseball batting. *Frontiers in Psychology*, *8*, 317328.
Gürbüz, E., & Taş, M. (2025). Investigation on the effect of virtual reality training on reaction time in football goalkeepers. *Turkish Journal of Sport and Exercise*, *27*(1), 112–117.
Handa, M., Aul, E. G., & Bajaj, S. (2012). Immersive technology–uses, challenges and opportunities. *International Journal of Computing & Business Research*, *6*(2), 1–11.
Hayden, S. (2024, October 22). Top NFL QB credits VR training as a big part of his success, calling it a "Flight simulator" for football. *Road to VR*. https://www.roadtovr.com/nfl-qb-vr-training-jayden-daniels-football-flight-simulator/
Huang, Y., Churches, L., & Reilly, B. (2015, April). A case study on virtual reality American football training. In *Proceedings of the 2015 Virtual Reality International Conference* (pp. 1–5). Laval France.
LaValle, S. M. (2023). *Virtual Reality*. Cambridge University Press.
Le Noury, P., Polman, R., Maloney, M., & Gorman, A. (2022). A narrative review of the current state of extended reality technology and how it can be utilised in sport. *Sports Medicine*, *52*(7), 1473–1489.
Lewis, M. (2024). Virtual reality, augmented reality, and implications for the modern sports landscape. In *Evolution of the Modern Sports Fan* (pp. 244–258). Bloomsbury Publishing.
Liu, C., Wang, W., Liu, H., & Wang, J. (2022). Application of Hawk-Eye technology to sports events. In *2022 2nd International Conference on Information Technology and Contemporary Sports (TCS)* (pp. 1–5). Guangzhou, China: IEEE.
Liu, S., Wu, C., Xiao, S., Liu, Y., & Song, Y. (2024). Optimizing young tennis players' development: Exploring the impact of emerging technologies on training effectiveness and technical skills acquisition. *Plos One*, *19*(8), e0307882.
Marr, B. (2021). *Extended Reality in Practice: 100+ Amazing Ways Virtual, Augmented and Mixed Reality Are Changing Business and Society*. John Wiley & Sons.
McClure, C., & Schofield, D. (2019). Running virtual: The effect of virtual reality on exercise. *Journal of Human Sport & Exercise*, 15, 861–870.
Meneses, J., Souto, L., Sil, S., Silva, T., & Au-Yong-Oliveira, M. (2022, April). Sports streaming: Transition from TV to digital: An analysis of traditional and new media companies in the online environment. In *World Conference on Information Systems and Technologies* (pp. 220–232). Cham: Springer International Publishing.
Michalski, S. C., Szpak, A., & Loetscher, T. (2019). Using virtual environments to improve real-world motor skills in sports: A systematic review. *Frontiers in Psychology*, *10*, 2159.
Pastel, S., Klenk, F., Bürger, D., Heilmann, F., & Witte, K. (2025). Reliability and validity of a self-developed virtual reality-based test battery for assessing motor skills in sports performance. *Scientific Reports*, *15*(1), 6256.
Petri, K., Emmermacher, P., Danneberg, M., Masik, S., Eckardt, F., Weichelt, S., Bandow, N., & Witte, K. (2019). Training using virtual reality improves response behavior in karate kumite. *Sports Engineering*, *22*, 1–12.

Potwarka, L. R., Hall, P. A., Goebert, C., & Ayaz, H. (2022). Immersive technology and the virtual sport spectator experience. In Naraine, M. L., Hayduk, T., & Doyle, J. P. (Eds.), *The Routledge handbook of Digital Sport Management* (pp. 232–244). Routledge.

Renganayagalu, S. K., Mallam, S. C., & Nazir, S. (2021). Effectiveness of VR head mounted displays in professional training: A systematic review. *Technology, Knowledge and Learning*, *26*(4), 999–1041.

Richlan, F., Weiß, M., Kastner, P., & Braid, J. (2023). Virtual training, real effects: A narrative review on sports performance enhancement through interventions in virtual reality. *Frontiers in Psychology*, *14*, 1240790.

Santana, D. (2019, December 10). Augmented reality has its limits, but is here to stay in sports. *Front Office Sports*. https://frntofficesport.com/augmented-reality-growth-in-sports/

Schäfer, A., Reis, G., & Stricker, D. (2022, September). Controlling continuous locomotion in virtual reality with bare hands using hand gestures. In *International Conference on Virtual Reality and Mixed Reality* (pp. 191–205). Cham: Springer International Publishing.

Soltanabadi, S., Minoonejad, H., Bayattork, M., & Seyedahmadi, M. (2023). Effect of virtual reality and augmented reality training for injury prevention and accelerating rehabilitation of Anterior Cruciate Ligament Injury in Athletes: A scoping review. *Asian Journal of Sports Medicine*, *14*(4), 1–18.

Stern, J. (2024, February 12). A fake yellow line changed football forever. *The Atlantic*. https://www.theatlantic.com/technology/archive/2024/02/football-virtual-yellow-line-augmented-reality/677384/

Suh, A., & Prophet, J. (2018). The state of immersive technology research: A literature analysis. *Computers in Human Behavior*, *86*, 77–90.

Syed, T. A., Siddiqui, M. S., Abdullah, H. B., Jan, S., Namoun, A., Alzahrani, A., Nadeem, A., & Alkhodre, A. B. (2022). In-depth review of augmented reality: Tracking technologies, development tools, AR displays, collaborative AR, and security concerns. *Sensors*, *23*(1), 146.

Wood, G., Wright, D. J., Harris, D., Pal, A., Franklin, Z. C., & Vine, S. J. (2021). Testing the construct validity of a soccer-specific virtual reality simulator using novice, academy, and professional soccer players. *Virtual Reality*, *25*, 43–51.

Xie, B., Liu, H., Alghofaili, R., Zhang, Y., Jiang, Y., Lobo, F. D., Changyang, L., Huang, H., Akdere, M., Mousas, C., & Yu, L. F. (2021). A review on virtual reality skill training applications. *Frontiers in Virtual Reality*, *2*, 645153.

Young, J. (2020, November 1). Sports leagues are betting on augmented reality, as virtual courtside seats can't match the real thing. *CNBC*. https://www.cnbc.com/2020/11/01/tech-augmented-reality-sports-leagues-nba-mlb-nhl-profit-virtual-reality.html.

7

ESPORTS AND DIGITAL TECHNOLOGY

Armand Buzzelli, Artemisia Apostolopoulou, and John Clark

Introduction

Esports, defined as organized, competitive video gaming in which individuals or teams play specific titles under standardized rules and structures, often before live or online audiences and for prize money, sponsorship, or ranking points (Abeza et al., 2024), have transitioned from the margins of entertainment into the mainstream of the sports industry. Esports has professional leagues across the world with multimillion-dollar investments rivaling traditional sports. A Statista (2024) report estimated that the global esports market was estimated to be worth 3.75 billion US dollars in 2023 and that, by 2028, the market is anticipated to surpass 5.67 billion US dollars, reflecting a compound annual growth rate (CAGR) of nearly 8.7% over the forecast period.

From emerging applications that provide data-driven audience experiences (Kokkinakis et al., 2020) to biometric tracking and eye movement analysis (Wang et al., 2024), and implementing non-invasive brain stimulation (NIBS) into performance training (Zhuang et al., 2020), esports is a data-intensive and performance-driven environment driven by technological innovation. This performance culture mirrors the rigor found in elite physical sport. Despite these advancements, esports athletes deal with a number of issues, including burnout (Madden & Harteveld, 2021), exploitation, and a lack of regulatory and legal protections (Dominteanu et al., 2023).

This chapter addresses the timely and critical issue of performance and ethics in esports, a topic with increasing relevance to players, fans, coaches, and scholars alike. As the industry continues to scale at an unprecedented speed, stakeholders need to consider questions around fairness, health, safety, and industry sustainability. The chapter proceeds through a historical overview of esports, explores key performance and ethical concepts, current trends, and case studies, and offers practical implications. The purpose of this chapter is to examine how the performance-oriented and decentralized nature of esports

DOI: 10.4324/9781003588405-9

both enhances and complicates its ethical landscape. By the end of this chapter, readers will have a clearer understanding of the landscape of competitive esports, including its structure, governance challenges, and cultural significance. They will be able to identify major ethical issues facing esports stakeholders, such as labor rights and performance enhancement. Finally, readers will get insights into how teams, leagues, and governing bodies are responding, or in some cases, failing to respond to challenges in the esports ecosystem.

Background

The early roots of esports can be traced back to video game-like simulation experiments such as the "Cathode Ray Tube Amusement Device" in the late 1940s and 1950s (Kent, 2001), followed by the development of *Spacewar!* at MIT in 1962 (Donovan, 2010). In 1972, an "Intergalactic Spacewar! Olympics" was held at Stanford University, and is widely recognized as the first known organized gaming competition (Borowy et al., 2013). By 1999, Mat Bettinson helped coin the term "esport" in association with the launch of the Online Gamers Association (Gestalt, 1999), serving as a label for the emerging trend of competitive gaming.

The COVID-19 pandemic accelerated the global visibility and engagement of the esports industry (Fakazli, 2020), solidifying its position as a worldwide phenomenon despite short-term economic challenges. One of the impacts of digitization (the conversion of content and services into online interactive digital formats) was that larger audiences were finding new forms of digital entertainment. Government-imposed measures that included in-home lockdowns, social distancing, and masking mandates changed how people interacted, socialized, and entertained themselves. With countries facing the challenges of navigating a global pandemic, the sports world also faced a series of crises (Türkmen & Özsarı, 2020). Traditional sports were briefly put on hold, and fans of sports competitions had limited options for consuming the games they loved. While people were homebound, participation in online games exploded (King et al., 2020).

Esports proved to be one of the few options for sports competitions that could be held remotely during the height of the pandemic. Despite mass cancellations of in-person esports local area network (LAN) events and the subsequent negative impact on those revenue streams, the digital nature of esports allowed the industry to focus on streaming and growing its broadcasting revenue (Fakazlı, 2020). People flocked to digital entertainment via Twitch, YouTube, and Facebook Gaming streaming platforms. According to Statista (2020), a record 22.7 million peak active users streamed games on Twitch during March 2020. This increased attention is one of the factors that spurred a wave of private investment dollars to the industry.

Despite explosive growth in viewership and financial investment, governance remains a major hurdle for the esports industry. The lack of a unified regulatory framework creates inconsistencies in how competitions are run, leading to questions about fairness, player rights, and performance standards. This chapter will explore these key concepts and other current issues in esports.

Topical Exploration and Key Concepts

Evolution of Esports Performance and Opportunities for Competition

While debates have raged for years on the merits of calling esports a sport (Rosell Llorens, 2017; Thiel & John, 2018), there is little argument that both the opportunities for organized competition and the attention esports have attracted globally have grown exponentially over the past decade (Block & Haack, 2021; Yadav & Utreja, 2024). Research has found that there is a significant incongruity in the terminology used to classify high-level esports samples and inconsistent justifications for what makes a sample elite (Poulus et al., 2024). Motivations for collegiate esports participants resemble athletes who compete in largely individual competitions, such as cyclists or runners, whereby athletic goals are intrinsically motivated and are separate from other important self-identities (Buzzelli & Draper, 2021). Professional esports coaches and athletes share similar perspectives to traditional sports athletes and coaches on the skill requirements and challenges that participants face (Railsback & Caporusso, 2019). Similar to traditional sports, the equipment and practice methods used in esports have improved over time, while the popularity and structure of various game titles continue to evolve alongside technological advancements (Bialecki et al., 2024).

Take the sport of golf as an example of technological advancements in equipment: the old wooden drivers used by Ben Hogan and his contemporaries pale in comparison to the capabilities of lightweight titanium woods with oversized heads used by Tiger Woods and Scottie Scheffler. As advancements in equipment, athlete training, and nutrition make their way into the sport, course records continue to be broken. The same logic applies to gaming as dramatic improvements are made by developers and engineers to the power and responsiveness of gaming computers and peripherals, such as modern gaming keyboards and mice. Unlike the golf example, a newly developed popular game may have a unique interface and thus be categorized differently across a broad spectrum of other game titles. This phenomenon makes it difficult to compare players from game title to game title and from era to era.

Hedlund and his colleagues (2021) categorized games across the following categories: multiplayer online battle arena (MOBA), first-person shooter (FPS), fighting games, sports games, racing games, real-time strategy (RTS), hybrid games, card games, and mobile-only games. Players most motivated by competition tend to gravitate more toward FPS and sports games, while there are large professional scenes across most of the aforementioned categories of games (Hedlund, 2021). Games with cartoonish or realistic violence have sparked debates about their impact on player behavior and public perception, particularly concerning esports' inclusion in mainstream sporting events. A lingering debate ensues about which games are appropriate for mainstream audiences, with some questioning whether games rife with violent content should be part of major esports organizations and sponsored interscholastically (Eisenhauer, 2023).

A variety of professional, collegiate, and K-12 leagues have sprung up over the years in a market that has a growing demand for governance (e.g., Overwatch League, League of Legends Championship Series, National Association of Collegiate Esports, High School Esports League). Many organizations are vying for control over interscholastic esports competitions. In addition to challenges deciding on which games to sponsor, interscholastic (collegiate and K-12) esports organizers need to cooperate with any restrictions set forth by game developers. Both for-profit and non-profit organizations, including the Network of Academic and Scholastic Esports Federations (NASEF), PlayVS, High School Esports League (HSEL), and the Interstate Scholastic Esports Alliance (ISEA), provide opportunities for high-school and middle-school competitions across the country. In the United States, in states in which traditional sport organizations do not sponsor esports, organizations such as the Pennsylvania Interscholastic Esports Association (PIEA) attempt to govern interscholastic play across their State.

Like high school, collegiate esports are not centralized in the same fashion as traditional athletics. The National Collegiate Athletic Association (NCAA) has not become involved in the governance of esports in any capacity at the time of this publication. The National Esports Collegiate Conference (NECC) and the National Association of Collegiate Esports (NACE) are two of the major players in terms of recognition and number of teams in the collegiate scene. Additionally, a growing number of esports programs at colleges and universities organize competitions across traditional athletic conferences, such as the Big Ten Esports Conference. Various organizers, including developers and third-party organizations, organize tournaments such as *College League of Legends* [https://lol.fandom.com/wiki/College_League_of_Legends], *Collegiate Rocket League* [https://rocketleagueesports.com], and the *Red Bull Campus Clutch* (*Valorant*) [https://www.redbull.com/int-en/event-series/campus-clutch]. While unofficial rankings are widely published for collegiate esports programs (e.g., *Colleague.GG College Esports Rankings* [https://colleague.gg]), no single governing body is widely recognized as the official organization for collegiate rankings.

As mentioned previously, professional esports have grown exponentially in popularity in recent years. As part of a natural evolution, the industry has expanded to offer an array of career opportunities, with professional players being among the most prominent (Johnson & Woodcock, 2021). In top professional organizations, players compete full-time, often living in team houses to maintain a competitive environment where they are supported by staff such as fitness trainers and chefs to ensure peak physical and mental performance (Ibrahim & Lewis, 2019). Esports is both a career and a lifestyle for high-level participants. In contrast to traditional sports, many esports professionals balance rigorous competitive schedules with content creation responsibilities, livestreaming, and personal branding (Witkowski & Manning, 2019).

Popular esports titles like League of Legends, CS:GO, Dota 2, and Fortnite offer significant prize pools, which have helped bring excitement and viewership to major events. The esports tournament that has offered the greatest overall prize pool was "The International 2021" Dota 2 tournament, with a total prize pool of over $40 million (Statista, 2024). Spectators of mega tournaments such as "The International" watch competitions in person in arenas, while even larger audiences watch livestreams of the event online. Esports players are generally younger compared to professionals in traditional sports, with pre-teens and early teenagers playing professionally. According to Luo et al. (2024, p. 10), "They face the threat of a shorter peak period in their careers. If they cannot achieve results quickly, they can easily be replaced by the younger generation, entering a phase of professional decline."

Professional esports is not only growing in popularity but also in both economic and cultural impact (Pasaribu, 2023). The increased viewership and prize pools (Ward & Harmon, 2019) have enabled it to become a viable career possibility for elite players. The growth in the industry helps contribute to a broader ecosystem, which boosts innovation and change.

Esports Athlete Development and Support Systems

The NASEF has developed a version of an esports ecosystem, representing transferable skill development and career opportunities within the scholastic esports industry. The domains of the NASEF Esports Ecosystem include the following:

- Strategists who work as analysts, coaches, and theory crafters
- Content creators such as streamers, shoutcasters, and journalists
- Organizers who manage events, IT support, and general operations
- Entrepreneurs who engage in marketing, business development, and corporate sponsorships.

At the core of this ecosystem are the players and games, and surrounding the core are the various functions that support and sustain the industry, reflecting the breadth of career pathways available (NASEF, n.d.).

Similar to traditional athletics, many scholastic esports coaches are employees of the schools and may be working at the school in a more primary capacity as a teacher, counselor, or some other form of student support personnel (Farber & Merchant, 2025). Esports coaches provide important life skill development, including leadership, teamwork, and communication skills (Farber & Merchant, 2025). Moreover, in addition to life skills, esports participants develop career skills that may benefit them in future employment endeavors, such as STEM skills, communication, teamwork, concentration, and quick decision-making (Rothwell & Shaffer, 2019).

According to the NACE, 47% of collegiate esports programs are housed in Club Sport or Student Affairs umbrellas within a university, 40% are overseen by Athletic Departments, while the remaining 13% of member schools house their esports program elsewhere on campus, in areas such as admissions or academic departments (Bauer-Wolf, 2019). As these programs vary broadly, so too does the amount of support esports athletes receive from health and wellness resources on their respective campuses.

Professional esports players navigate dual roles in their careers as both esports competitors and streamers. This dynamic adds layers of psychological and professional strain that go beyond the typical stressors that are part of competition. As Seula et al. (2018) describe in their case study of Brazilian League of Legends pro-player "brTT," these players must carefully manage distinct personas for the championship stage and the live-streaming arena – both requiring continuous performance, attention, and composure. While the professional setting demands coordination with a team and more restraint, the streaming environment is more informal, but it demands an entertaining performance, audience engagement, and a great deal of stamina to succeed. This multifaceted identity may help attract fans and sponsorship opportunities.

Sports and esports are unique in that consumers (fans) have emotional ties to the product in a way that is unique to nearly any other industry. One study found that the strength of fan identity was a significant predictor of esports viewing frequency and engagement within esports fandoms (Barney & Pennington, 2023). Fans are active participants, and their emotional investment drives key revenue streams for leagues and franchises, including ticket sales, merchandise, sponsorship deals, and key digital media metrics. At the same time, the pressure to please fans may also amplify the mental strain on players. Coaches may be called upon to provide elite esports athletes with the mental strategies to help regulate their emotions and remain mindful in the presence of "tilt" (Poulus et al., 2024).

Physical and Mental Well-Being of Esports Athletes

Despite the rise of coaches and support personnel, studies have suggested that a high volume of playing video games may result in undesirable physical and mental health effects. The esports community has seen a rise in cyberbullying incidents involving harassment, humiliation, blackmailing, and slandering (Polat & Şenbiçer, 2022). Continuous exposure to this type of harassment can lead to burnout and mental health issues for players. Esports organizations must develop policies and support systems to protect athletes from cyberbullying while cultivating a positive culture.

Tseng (2020) analyzed the Entertainment Software Association's (ESA) announcement of four universal principles for esports: safety and well-being, integrity and fair play, respect and diversity, and positive and enriching gameplay. He argued that these principles could form the basis for a universal code of conduct for esports governance that would help mitigate issues in toxic behaviors that threaten player wellness and inclusivity. Several organizations have implemented practical tools such as honor systems and in-game reporting and banning policies to help reduce toxic behavior. Blizzard introduced a player endorsement system that is a robust disciplinary program that combines positive reinforcement (loot boxes, public recognition) with social incentives (increased match access). The system reduced toxic behavior by over 25% and helped reshape norms and expectations in the community (Tomkinson & van den Ende, 2022).

Additionally, it is not uncommon for esports participants to experience burnout due to the demands of their careers while trying to balance other aspects of their lives (Madden & Harteveld, 2021). Long hours of both team practice and solo training often lead to social isolation, making it difficult to maintain any semblance of a healthy work–life balance. Despite many esports titles being team-based in terms of competition, teams are typically small; games like Rocket League often consist of three-player teams, and team activities are typically virtual.

While much of the research related to esports players is focused on mental health, the physical aspects of participation also need to be addressed to support player wellness. Professional esports participants cite a litany of challenges preventing them from reaching peak performance, including long practice hours, player attitudes, mental stress, and other health concerns such as back and wrist injuries (Sabtan, 2022). Abnormal sitting postures, repetitive movements, prolonged screen exposure, extended gaming hours, and a sedentary lifestyle can contribute to various medical issues affecting the musculoskeletal, ophthalmological, neurological, and metabolic systems (Rossoni et al., 2023).

Along similar lines, DiFrancisco-Donoghue and her colleagues (2019) reported that nearly half of gamers experience eye fatigue, neck and back pain, and wrist and hand pain. Furthermore, participation in esports has been linked to intensive training schedules, lack of recovery time, sleep deprivation, and the resulting mental and physical exhaustion that contribute to the widely reported burnout experienced by esports athletes. Ahn and Kim (2024) studied burnout among professional esports athletes and cautioned that "due to the characteristics of esports, it is necessary to pay attention to the training schedule that lasts from afternoon to dawn" (p. 5).

Research has shown that physical activity can enhance player performance in terms of accuracy, reaction time, and cognitive flexibility (McNulty et al., 2023). Furthermore, De Las Heras et al. (2020) suggest that pre-game workouts can improve performance by up to 75% in precision tasks. These studies offer compelling evidence for additional investment in conditioning personnel and programming by esports properties similar to those of traditional sports organizations.

Nevertheless, it is rare to find organizations across the esports industry offering structured physical activity routines. A large-scale study by Nagorsky and Wiemeyer (2020) involving over 1,800 players found that, although many players acknowledge the benefits of fitness, few regularly engage in esports-specific physical training. Their research proposed an integrative performance model that blended game science and sports science, and identified six key components of esports performance: sensorimotor coordination, cognitive-tactical skills, physical conditioning, emotional stability, social communication, and media literacy. Their research emphasized that esports training should be game-specific and that physical and psychological preparation should be core elements of athlete development.

Athlete Rights and Ethical Standards in Esports Competition

The career development of esports athletes does not follow traditional linear models seen in conventional sports. Esports athletes typically start younger than traditional athletes, further compounding the unpredictability of the industry. Meng-Lewis and her colleagues (2021) applied the Chaos Theory of Careers (CTC) to illustrate how esports careers are characterized by constant change, complexity, and unpredictability, making long-term career planning challenging for young players and their families. Many players retire at a young age citing physical and mental fatigue, as well as shifts in the competitive landscapes of their games of expertise.

In addition to market instability and career uncertainty, the lack of structured career planning and educational continuity can leave players with limited opportunities once they step away from professional competition (Hong, 2025). Unlike traditional sports, the esports ecosystem does not offer a clear pathway for players to advance from amateur competition to the Tier 1 global professional rank due in large part to the lack of semi-professional structured opportunities (Partin, 2019). In traditional sports, athletes usually go to college before turning professional, but in esports this path can be reversed. Players who are past their competitive prime can further their education and earn scholarships to compete in collegiate esports. This "esports backwards collegiate pipeline" is part of the Esports Pathway to Pro diagram developed by Jenny (2024), which is illustrated in Figure 7.1.

As esports continues to evolve into a global industry, player rights must be addressed. It is common for team owners or tournament organizers to be unable to pay their players adequately or use coercive tactics to underpay them (Ronquillo, 2018). The unique structure of the esports workforce, which consists primarily of short-term contracts (Dominteanu et al., 2023), contributes to career instability. Addressing these challenges requires a more structured approach to player protection, labor rights, and long-term career development within the industry.

Esports professionals also face significant legal and labor-related issues. Unlike athletes in the four major professional sports leagues in the United States (MLB, NBA, NFL, and NHL), esports leagues typically don't have player unions or collective bargaining agreements. They may enter the professional scene at a very young age and sign contracts without expert legal representation (Cisneros, 2022). Furthermore, the global nature of esports means that labor laws and regulations vary widely across countries. This further complicates efforts to develop standard player protections. While some countries have implemented esports-specific regulations, the industry lacks universal governance, which impacts player rights and contracts (Çakar & Yiğit, 2019). In response to these challenges, Patel (2019) argues that unionization could empower esports players, grant them collective bargaining power to negotiate fair contracts, improve working conditions, and improve job security. Future efforts should focus on developing regulations that support player welfare and career development.

Kelly and her colleagues (2021) developed a best practice framework for addressing integrity and well-being issues through esports governance. The core pillars of their process are as follows:

- Representation of all stakeholders in the policy development process
- Engagement of youth as agents with rights
- Monitoring of stakeholder interactions, policy development, and policy implementation (e.g., an international governing body or governments monitoring national governing bodies)
- Ongoing evaluation of the effectiveness of the policy implementation and the dynamic evolution of it
- Inclusion of fluid and complex, networked interactions among stakeholders.

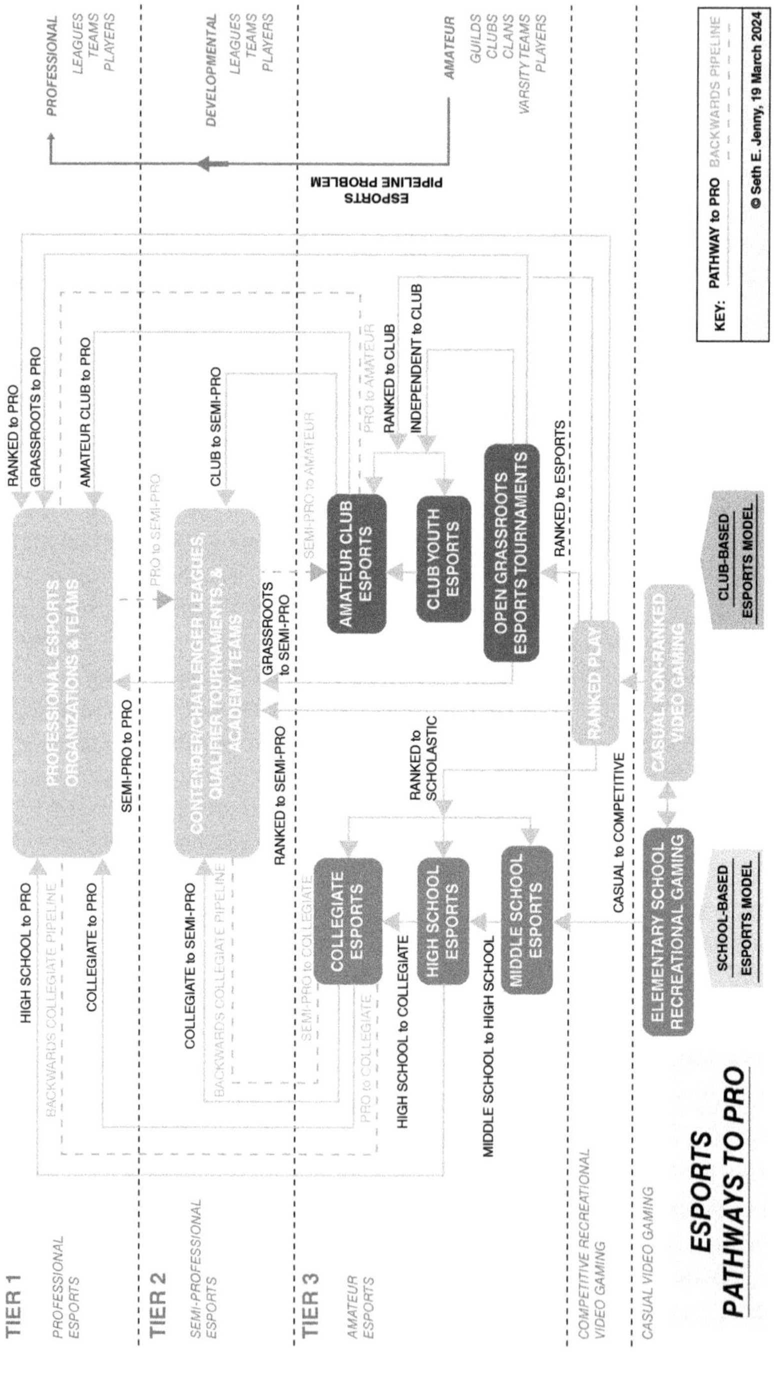

FIGURE 7.1 Esports Pathways to Pro.

Reprinted with permission from Seth E. Jenny (2024).

The framework offers quality advice on the development of policy and governance in esports. These scholars suggest that there is a

> Complexity of wellbeing issues linked to esports due to its unique, digital, and transnational context, and largely minor audiences support the establishment of national and international governance agencies to monitor, sanction, and advise about the policy-making processes of subordinate bodies.
>
> *(Kelly et al., 2021, p. 12)*

This holistic framework may provide a roadmap for esports governance.

Audience values are an important element to consider when building governance. Spectators judge esports harshly when the integrity of a competition is breached. Even though fan perceptions of cheating in esports vary, any questions about participant integrity can erode public trust (Johnson & Abarbanel, 2022). This perception of fairness is an important factor in the sustainability of the industry moving forward and must be incorporated into governance plans.

In closing, ethical frameworks must go beyond reactive policies and embrace sustainable player-focused governance. In the fast-paced world of esports, ethics cannot be treated simply as a theoretical concern; instead, they need to be at the heart of how business decisions are made across the industry (Karhulahti & Kimppa, 2018). Current models such as the one at Riot Games blend commercial strategy with community building but lack ethical foresight (Chee & Karhulahti, 2020). As an industry, governance needs to focus on future-proof ethical frameworks that balance power among publishers, teams, and players.

Practice in Action

Perspectives on Governance from the National Esports Collegiate Conference

Jacob VanRyn, Commissioner of the NECC, offers a pragmatic approach to the challenges of governing collegiate esports. "Our goal is not necessarily to be the NCAA," VanRyn says, "but there are some components of the NCAA we have tried to mirror in many ways to provide our membership - those include things like minimum requirements for student-athletes to be eligible to compete, updated rule-sets, and bylaws documents" (J. VanRyn, personal communication, January 23, 2025).

Since nearly all NECC competitions occur virtually, player verification is an important part of maintaining the integrity of competition. VanRyn, who has previous experience in traditional sports, notes the difference: "There is no field, or gym where players line up next to or opposite one another for competitions. This presents challenges like player verification and ensuring that every player who is competing with us is representing who they are." VanRyn also acknowledges that ethical challenges, including smurfing, cheating tools, and player misconduct, are concerns for governance:

> We have a competition committee that meets regularly whenever any sort of issue comes up to complete a full investigation on an as-needed basis. We never want to deal with a claim of cheating or inappropriate behavior, but every time we have to go through a full investigation it makes us a better organization overall.
>
> *(J. VanRyn, personal communication, January 23, 2025)*

Looking to the future, VanRyn envisions a more holistic role for the NECC. The league has partnered with FitGMR to promote physical and mental health among players. "Not enough schools take advantage of it yet," VanRyn says, "but we are emphasizing it moving forward" (J. VanRyn, personal communication, January 23, 2025).

Trends and Current Issues

Esports and "Unhealthy Behavior" Sponsorship

Esports organizations continue to wrestle with the ethical implications of certain types of sponsorships that are common across the industry. Brands in industries such as energy drinks, gambling, tobacco, unhealthy snack foods, cryptocurrency, and NFTs have increasingly targeted esports audiences that are heavily populated with teenagers and younger adults. While energy drinks are now embedded in the popular culture of esports, they have significant health risks.

Lopez Frias (2021) compares energy drink marketing to historical tobacco and alcohol marketing in traditional sports and suggests that the drinks function as a form of "neuro-doping." He argues that, "Targeting the gaming community is the most effective, lucrative, and direct way to reach children and teenagers … the gaming market is the 'wild west for advertising'" (Lopez Frias, 2021, p. 240). Furthermore, alcohol sponsorship in esports has been linked to increased awareness, preference, and consumption while gaming among heavy gamers (Kelly & Van der Leij, 2021). That particular study, which was based on nearly 1,000 Australian gamers aged 16–34, reported that younger, heavy users are more susceptible to alcohol brand messaging in esports content.

Gambling is another "unhealthy behavior" that may be stimulated by esports sponsorship agreements. According to Greer et al. (2021, p. 442), "Esports bettors experience significantly higher levels of gambling problems and gambling-related harms compared to those who only gamble on traditional gambling activities." That same study found that esports bettors tended to be younger than their traditional sports betting counterparts, highlighting the risky nature of gambling-related sponsorships that may create a pathway for underage users to engage in gambling. Given all this evidence, the esports industry would be best served by using ethical standards to screen potential sponsorship agreements, prioritizing the well-being of its younger audiences, athletes, and fans.

Maximizing Gaming Technologies for Societal Impact

Gamification is a flexible and scalable tool that meets people where they frequently are: on their phones, mobile devices, or wearable technology. Game design elements can be implemented into public health campaigns, physical activity adherence, and mental wellness programs. Tran and his colleagues (2022) reported mixed but promising results for mobile health apps using a combination of gamification and financial incentives with patients. Other studies have offered evidence in favor of gamification's positive effect in primary prevention efforts (Tesi et al., 2023), as well as the use of video games in psychological and physical therapy with positive outcomes (Primack et al., 2012).

Video game technologies, including VR, simulation, and gaming controllers, are transforming how surgeons train and perform complex procedures. In 2024, surgeons in Switzerland performed an endoscopy on a pig over 6,000 miles away using a PlayStation controller. According to McNally (2024), "This remarkable leap in health technology could lead to operations being performed in remote locations where the necessary world-class expertise might not necessarily be available." Another example of the medical field using gaming technology for education and innovation includes the use of virtual reality and simulation to enhance technical skill acquisition and safety through repeatable training environments (Beqari & Seymour, 2021). Virtual reality specifically offers a great deal of potential for medical innovation, supporting patients throughout the surgical care pathway from preoperative education to postoperative rehabilitation (Gendia et al., 2023), and through remote medical education, particularly under the COVID-related restrictions (Pallavicini et al., 2022).

Wellness Departments in Higher Education and Esports

Esports programs may begin to find a new home on college campuses that could better position collegiate esports athletes to deal with the wellness challenges outlined in this chapter. Wellness divisions have begun to emerge at institutions across the country, a reflection of higher education's shift toward a more holistic approach to student development. As colleges and universities combat the student mental health crisis, many are hiring Chief Wellness Officers (CWOs) to centralize initiatives and programming that may include areas of student life such as campus recreation, student health services, counseling services, and spiritual or religious programs. "The time is now for enacting change to create, improve, or sustain cultures of wellness within campus communities" (Amaya et al., 2019, p. 27), and wellness divisions provide a better framework to do so.

Esports and Wellness at Robert Morris University

Robert Morris University (RMU) recently made an executive decision to place esports into its newly developed wellness division and under the leadership of its CWO Dr. Jason Miller. This decision reflects a growing trend in higher education where wellness is no longer viewed as a standalone service, but is strategically incorporated into the overall student experience. Miller, who also oversees Campus Recreation, views esports as more than just competition:

> Collegiate esports programs provide avenues for students to engage in a variety of experiences and acquire new skills that directly connect to various aspects of wellness, allowing for holistic development and reinforcement of their health and well-being.
>
> *(J. Miller, personal communication, March 25, 2025)*

While acknowledging challenges to their physical wellness related to the sedentary nature of esports coupled with nighttime competitions that may impact the quality of sleep that his players get, Miller feels that esports is now better positioned for success:

> With an embedded focus on health and wellbeing, we anticipate greater positive outcomes for our esports participants, including academic performance, time/priority management, physical wellness, and advanced social-emotional skills.
>
> *(J. Miller, personal communication, March 25, 2025)*

RMU Esports Director Travis Bolner shares his philosophy: "Collegiate esports often fixates on the wrong aspects," he says. "If a student logs ten hours of gameplay, I want to know if that replaced study time. Did it impact their sleep? Collegiate esports should be education first, esports second." To help RMU esports students maintain balance, Bolner enforces strict GPA and attendance policies, referring students to campus wellness professionals when additional support is needed. Bolner also notes that being housed within the wellness division has provided students with immediate access to resources that help them thrive.

> We want our students to prosper in and out of gameplay, he says. And we pride ourselves on being able to offer them the tools they need to do that in a healthy, sustainable way.
>
> *(T. Bolner, personal communication, March 24, 2025)*

Summary

This chapter explored the dynamic relationship between performance and ethics in the growing world of esports. Drafting off of its explosive growth during the COVID-19 pandemic, the industry continues to evolve in a variety of ways. With advancements in technology and training methods, esports players operate in a sophisticated environment. They are faced with pressures to both perform in high-level competition and become influencer personalities that are marketable enough to generate revenue for themselves and their organizations. The pressure that accompanies these roles when combined with inconsistencies across industry governance has led to player burnout and turnover.

To support long-term success, stakeholders must prioritize the well-being of esports athletes. This can be done through a combination of wellness initiatives, career support, and ethical governance. Trends like controversial sponsorships, the integration of gaming technologies into healthcare, and the emergence of wellness divisions in institutions of higher education illustrate both the challenges and opportunities for esports and gaming.

Key Resources

- **Esports Integrity Commission (ESIC)** – https://esic.gg
- **Esports Research Network** – https://esportsresearch.net/literature/
- **International Esports Federation (IESF)** – https://iesf.org
- **National Esports Collegiate Conference (NECC)** – https://necc.gg
- **Twitch** – https://www.twitch.tv

References

Abeza, G., Wanless, E., & Mamo, Y. (2024). Emerging developments in sport management. In P. Pedersen & L. Thibault (Eds.), *Contemporary Sport Management* (pp. 283–308). HK.

Ahn, H., & Kim, I. (2024, May). An exploratory study on the conceptualization of burnout among the professional esports athletes: Focused on League of Legends Champions Korea League. *Healthcare*, *12*(11), 1127. https://doi.org/10.3390/healthcare12111127

Amaya, M., Donegan, T., Conner, D., Edwards, J., & Gipson, C. (2019). Creating a culture of wellness: A call to action for higher education, igniting change in academic institutions. *Building Healthy Academic Communities Journal*, *3*(2), 27–40.

Barney, J., & Pennington, N. (2023). An exploration of esports fan identity, engagement practices, and motives. *Telematics and Informatics Reports*, *11*, 100081. https://doi.org/10.1016/j.teler.2023.100081

Bauer-Wolf, J. (2019, February 12). A new frontier in college athletics: Video games. Retrieved December 2, 2024, from https://www.insidehighered.com/news/2019/02/12/new-frontiercollege-athletics-video-games

Beqari, J., & Seymour, N. E. (2021). Application of technology to educational needs in surgery. *Journal of Surgical Oncology*, *124*(2), 181–192.

Białecki, A., Michalak, B., & Gajewski, J. (2024). Esports training, periodization, and software—A scoping review. *Applied Sciences*, *14*(22), 10354. https://doi.org/10.3390/app142210354

Block, S., & Haack, F. (2021). Esports: A new industry. EDP sciences. In *SHS Web of Conferences* (p. 92). https://doi.org/10.1051/shsconf/20219204002

Borowy, M., Jin, D. Y., & Pluda, A. (2013). Pioneering eSport: The experience economy and the marketing of early 1980s arcade gaming contests. *International Journal of Communication*, *7*, 1–21.

Buzzelli, A., & Draper, J. (2021). Are they athletes? A self-assessment of athletic identity measurement and perceived benefits of collegiate Esports participants. *Recreational Sports Journal*, *45*(2), 117–130.

Çakar, D. B., & Yiğit, M. H. (2019). eSports eSports regulations and problematics. *The Journal of Eurasia Sport Sciences and Medicine*, *1*(3), 123–137.

Chee, F., & Karhulahti, V. M. (2020). The ethical and political contours of institutional promotion in esports: From precariat models to sustainable practices. *Human Technology: An Interdisciplinary Journal on Humans in ICT Environments*, *16*(2), 200–226.

Cisneros, J. (2022). Leveling the e-sports playing field: An argument in favor of government regulation to ensure fair player contracts for young professional gamers in e-sports. *California Western Law Review*, *58*(2), 5

De Las Heras, B., Li, O., Rodrigues, L., Nepveu, J. F., & Roig, M. (2020). Exercise improves video game performance—A win-win situation. *Medicine & Science in Sports & Exercise*, *52*(7), 1595–1602. https://doi.org/10.1249/MSS.0000000000002277

DiFrancisco-Donoghue, J., Balentine, J., Schmidt, G., & Zwibel, H. (2019). Managing the health of the eSport athlete: An integrated health management model. *BMJ Open Sport & Exercise Medicine*, *5*(1), e000467. https://doi.org/10.1136/bmjsem-2018-000467

Dominteanu, T., Smîdu, N., Voinea, A., Dinciu, C. C., Porfireanu, M. C., & Iacobini, A. (2023). Player exploitation in Esports: Organizational policies and business strategies. In *Proceedings of the International Conference on Business Excellence*, Bucharest, Romania (Vol. 17, No. 1, pp. 1413–1420). Sciendo.

Donovan, T. (2010). *Replay: The History of Video Games*. Yellow Ant.

Eisenhauer, J. G. (2023). Collegiate esports and nonviolence. *Journal of Higher Education Management*, *38*(1), 153–166.

eSports Earnings. (2024, December 13). Leading esports tournaments worldwide as of December 2024, ranked by overall prize pool (in million U.S. dollars) [Graph]. *Statista*. Retrieved January 2, 2025, from https://www-statista-com.reddog.rmu.edu/statistics/517940/leading-esports-tournamets-worldwide-by-prize-pool/

Fakazlı, A. E. (2020). The effect of the COVID-19 pandemic on digital games and eSports. *International Journal of Sport Culture and Science*, *8*(4), 335–344.

Farber, M., & Merchant, W. (2025). Cultivating Esportspersonship: How scholastic esports coaches perceive facilitating positive youth development. *Simulation & Gaming*, *56*(2), 155–177. 10468781241299131.

Gendia, A., Rehman, M., Cota, A., Gilbert, J., & Clark, J. (2023). Can virtual reality technology be considered as a part of the surgical care pathway? *The Annals of the Royal College of Surgeons of England*, *105*(1), 2–6.

Gestalt. (1999, December 13). The OGA: What the hell is it? http://www.eurogamer.net/articles/oga

Ghoshal, A. (2019). Ethics in esports. *Gaming Law Review*, *23*(5), 338–343.

Greer, N., Rockloff, M. J., Russell, A. M., & Lole, L. (2021). Are esports bettors a new generation of harmed gamblers? A comparison with sports bettors on gambling involvement, problems, and harm. *Journal of Behavioral Addictions*, *10*(3), 435–446.

Hedlund, D. P. (2021). A typology of Esport players. *Journal of Global Sport Management*, *8*(2), 460–477. https://doi.org/10.1080/24704067.2021.1871858

Hedlund, D. P., Fried, G., & Smith, R. C. (2021). *Esports Business Management*. Human Kinetics, Inc.

Hong, H. J. (2025). Exploring stakeholders' perspectives on esports players' wellbeing and career prospects. *Computers in Human Behavior Reports*, *17*, 100589. https://doi.org/10.1016/j.chbr.2025.100589.

Ibrahim, F., & Lewis, L. (2019). Just a game: Adoption of franchising in esports. *Franchise Law Journal*, *39*, 371.

Jenny, S. E. (2024, March 19). *Research+*. https://www.sethjenny.com/research

Johnson, M. R., & Abarbanel, B. (2022). Ethical judgments of esports spectators regarding cheating in competition. *Convergence*, *28*(6), 1699–1718.

Johnson, M. R., & Woodcock, J. (2021). Work, play, and precariousness: An overview of the labour ecosystem of esports. *Media, Culture & Society*, *43*(8), 1449–1465.

Karhulahti, V. M., & Kimppa, K. (2018). "Two queens and a pawn, please." An ethics for purchase, loot, and advantage design in esports. In *GamiFIN Conference 2018*, Pori, Finland, May 21–23, 2018 (pp. 115–122).

Kelly, S. J., Derrington, S., & Star, S. (2021). Governance challenges in esports: A best practice framework for addressing integrity and wellbeing issues. *International Journal of Sport Policy and Politics*. https://doi.org/10.1080/19406940.2021.1976812

Kelly, S. J., & Van der Leij, D. (2021). A new frontier: Alcohol sponsorship activation through esports. *Marketing Intelligence & Planning*, *39*(4), 533–558.

Kent, S. L. (2001). *The ultimate history of video games: From Pong to Pokémon and beyond: The story behind the craze that touched our lives and changed the world*. Three Rivers Press.

King, D. L., Delfabbro, P. H., Billieux, J., & Potenza, N. (2020). Problematic online gaming and COVID-19 pandemic. *Journal of Behavioral Addictions*, *9*(2), 184–186. https://doi.org/10.1556/2006.2020.00016

Kokkinakis, A. V., Demediuk, S., Nölle, I., Olarewaju, O., Patra, S., Robertson, J., … Block, F. (2020, June). Dax: Data-driven audience experiences in esports. In *Proceedings of the 2020 ACM International Conference on Interactive Media Experiences*. Barcelona, Spain (pp. 94–105).

LA Times. (March 29, 2020). Number of users on Twitch as a result of the coronavirus (COVID-19) pandemic worldwide as of March 2020 [Graph]. *Statista*. Retrieved December 17, 2024, from https://www-statista-com.reddog.rmu.edu/statistics/1108340/covid-twitch-users/, https://www.statista.com/statistics/1108340/covid-twitch-users/

Lopez Frias, F. J. (2021). "Fueling up" gamers. The ethics of marketing energy drinks to gamers. *Neuroethics*, *14*(Suppl 2), 239–249. https://doi.org/10.1007/s12152-020-09442-8

Luo, Y., Liu, L., Luo, X., & Li, S. (2024). From gamers to players: Chinese Generation Z's perspectives on minors dropping out of school to become professional esports players. *International Review for the Sociology of Sport*. https://doi.org/10.1177/10126902241293309

Madden, D., & Harteveld, C. (2021, May). "Constant Pressure of Having to Perform": Exploring player health concerns in esports. In *Proceedings of the 2021 CHI Conference on Human Factors in Computing Systems*. Yokohama, Japan (pp. 1–14).

Market.us. (2024, March 31). eSports market size worldwide in 2022 and 2023, with a forecast to 2032 (in billion U.S. dollars) [Graph]. *Statista*. Retrieved December 17, 2024, from https://www-statista-com.reddog.rmu.edu/statistics/1256162/global-esports-market-size/

McCarthy, M. (2021). Balancing esports: Protecting the next generation's superstars. *Journal of High Technology Law*, *22*, 448.

McNally, P. (2024). *Surgeons use PlayStation Controller to Perform Endoscopy 6,000 Miles Away – Good Job the PlayStation Network Was Still up at that Point*. Newstex.

McNulty, C., Jenny, S. E., Leis, O., Poulus, D., Sondergeld, P., & Nicholson, M. (2023). Physical exercise and performance in esports players: An initial systematic review. *Journal of Electronic Gaming and Esports*, *1*(1), 1–11.

Meng-Lewis, Y., Wong, D., Zhao, Y., & Lewis, G. (2021). Understanding complexity and dynamics in the career development of eSports athletes. *Sport Management Review*, *25*(1), 106–133. https://doi.org/10.1016/j.smr.2020.08.003

Nagorsky, E., & Wiemeyer, J. (2020). The structure of performance and training in esports. *PloS One*, *15*(8), e0237584.

NASEF. (n.d.). *Career Pathways*. https://www.nasef.org/career-pathways

Pallavicini, F., Pepe, A., Clerici, M., & Mantovani, F. (2022). Virtual reality applications in medicine during the COVID-19 pandemic: Systematic review. *JMIR Serious Games*, *10*(4), e35000.

Partin, W. (2019). The esports pipeline problem. *Polygon*, July, 11, 2019. https://www.polygon.com/features/2019/7/11/18632716/esports-amateur-pro-players-teams-talent-process.

Pasaribu, A. M. N. (2023). Electronic sports (e-sports): A global phenomenon affecting sports culture. *Riyadhoh: Jurnal Pendidikan Olahraga*, *6*(2), 190–194.

Patel, R. (2019). Esports, player positions, and the benefits of unionization. *Duke Law & Technology Review*, *18*, 232.

Polat, V., & Şenbiçer, Ş. (2022). Esports business and cyberbullying: Risks and precautions. In E. Öngün, N. Pembecioğlu, & U. Gündüz (Eds.), *Handbook of Research on Digital Citizenship and Management during Crises* (pp. 94–115). IGI Global Scientific Publishing.

Poulus, D. R., Sharpe, B. T., Jackman, P. C., Swann, C., & Bennett, K. J. (2024). Defining elite Esports athletes: A scoping review. *International Review of Sport and Exercise Psychology*, *1*(1), 1–36. https://doi.org/10.1080/1750984X.2024.2386531.

Primack, B. A., Carroll, M. V., McNamara, M., Klem, M. L., King, B., Rich, M., … Nayak, S. (2012). Role of video games in improving health-related outcomes: A systematic review. *American Journal of Preventive Medicine*, *42*(6), 630–638.

Railsback, D., & Caporusso, N. (2019). Investigating the human factors in eSports performance. In *Advances in Human Factors in Wearable Technologies and Game Design: Proceedings of the AHFE 2018 International Conferences on Human Factors and Wearable Technologies, and Human Factors in Game Design and Virtual Environments*, Held on July 21–25, 2018, in Loews Sapphire Falls Resort at Universal Studios, Orlando, Florida, USA (pp. 325–334). Springer International Publishing.

Ronquillo, J. (2018). The rise of esports: The current state of esports, its impacts on contract law, gambling, and intellectual property. *Intellectual Property & Technology Law Journal*, *23*, 81.

Rosell Llorens, M. (2017). eSport gaming: The rise of a new sports practice. *Sport, Ethics and Philosophy*, *11*(4), 464–476.

Rossoni, A., Vecchiato, M., Brugin, E., Tranchita, E., Adami, P. E., Bartesaghi, M., Cavarretta, E., & Palermi, S. (2023). The eSports medicine: Pre-participation screening and injuries management—An Update. *Sports*, *11*(2), 34. https://doi.org/10.3390/sports11020034

Rothwell, G., & Shaffer, M. (2019). eSports in K-12 and post-secondary schools. *Education Sciences*, *9*(2), 105.

Sabtan, B. (2022). *Human Factors in Esports: Investigating Performance Measures, Coaching Practices, and Stress Training in League of Legends*. UWSpace.

Schöber, T., & Stadtmann, G. (2022). The dark side of e-sports: An analysis of cheating, doping & match-fixing activities and their countermeasures. *International Journal of Esports*, *1*(1). https://www.ijesports.org/article/98/html

Seula, R., Amaro, M., & Fragoso, S. (2018). Play to win, profit & entertain: A study of double performance as athlete and streamer. In *Proceedings of DiGRA 2018 Conference: The Game is the Message*. Turin, Italy: Università di Torino.

Tesi, B., Donia, P., Covelli, A., Aloise, I., Renzi, E., Massimi, A., & Villari, P. (2023). "Health is a serious game": Gamification as a tool for primary prevention in Public Health. *European Journal of Public Health, 33*(Supplement_2), ckad160–032.

Thiel, A., & John, J. M. (2018). Is eSport a 'real'sport? Reflections on the spread of virtual competitions. *European Journal for Sport and Society, 15*(4), 311–315.

Tomkinson, S., & Van Den Ende, B. (2022). 'Thank you for your compliance': Overwatch as a disciplinary system. *Games and Culture, 17*(2), 198–218.

Tran, S., Smith, L., El-Den, S., & Carter, S. (2022). The use of gamification and incentives in mobile health apps to improve medication adherence: Scoping review. *JMIR mHealth and uHealth, 10*(2), e30671.

Tseng, Y. S. (2020). The principles of esports engagement: A universal code of conduct? *Journal of Intellectual Property Law, 27*, 209.

Türkmen, M. & Özsarı, A. (2020). Covid-19 salgını ve spor sektörüne etkileri. *International Journal of Sport Culture and Science, 8*(2), 55–67.

Wang, H., Yang, J., Hu, M., Tang, J., & Yu, W. (2024). A comparative analysis for eye movement characteristics between professional and non-professional players in FIFA eSports game. *Displays, 81*, 102599.

Ward, M. R., & Harmon, A. D. (2019). ESport superstars. *Journal of Sports Economics, 20*(8), 987–1013.

Witkowski, E., & Manning, J. (2019). Player power: Networked careers in esports and high-performance game livestreaming practices. *Convergence, 25*(5–6), 953–969.

Yadav, M. K., & Utreja, K. (2024). eSports on the rise: A statistical Odyssey unveiling the probability of growth. In *2024 International Conference on Electrical Electronics and Computing Technologies (ICEECT)* (Vol. 1, pp. 1–5). IEEE: Greater Noida, Ruhallapur, Uttar Pradesh.

Zhuang, W., Yin, K., Zi, Y., & Liu, Y. (2020). Non-invasive brain stimulation: Augmenting the training and performance potential in esports players. *Brain Sciences, 10*(7), 454.

8

DIGITAL TRANSFORMATION OF TRADITIONAL SPORTS

Cindy Lee and Gary Lhotsky

Introduction

As the industrial world has been evolving into a digital world (Parviainen et al., 2017), digital transformation becomes an important element for an organization's survival, growth, and development (Dašić, 2023), enabling companies to achieve greater efficiency and flexibility and optimize production processes (Chen et al., 2018; Queiroz et al., 2020). Digital transformation can be defined as the integration of digital technology into all aspects of business operation, which in turn leads to infrastructure changes in how the organization is operated (Vial, 2019). In other words, digital transformation involves integrating digital technologies (e.g., ticket systems, digital twins, cloud computing) into all aspects of business operations, enabling organizations to enhance efficiency, optimize decision-making, and remain competitive in an increasingly digital landscape. For example, the usage of digital twin technology has enhanced facility operations by creating a real-time virtual replica of physical assets, systems, or entire buildings, thus bringing benefits such as efficient predictive maintenance, energy optimization, and real-time monitoring and visualization, which can be useful for crowd control and emergency situations (Cespedes-Cubides & Jradi, 2024).

This digital transformation is also found in the sports industry, where digital technologies have changed the way sports organizations operate. Various technologies such as artificial intelligence (AI), big data analytics, and wearable technology impact all three user groups in sports, namely managers, athletes, and consumers (Frevel et al., 2022) in such areas as performance monitoring, fan engagement, and media delivery. More specifically, for example, AI-powered analytics now offer managers and coaches deeper insights into player performance, injury prevention, and game strategies, while wearable sensors track biometric data to optimize athletes' training and recovery. Meanwhile, digital broadcasting and streaming platforms have broadened

DOI: 10.4324/9781003588405-10

consumers' access to sports content worldwide. Furthermore, digital platforms and social media have reshaped fan engagement, creating direct communication channels and personalized content experiences. These digital innovations have not only influenced fan interaction but have also unlocked new revenue streams through virtual sponsorships and gamified engagement strategies.

Digital transformation has therefore significantly changed the way traditional sports are played, consumed, and managed. The following sections provide an overview of its history and background of the digital transformation of traditional sports, key areas of impact, key trends, and a summary.

Background

The integration of technology into traditional sports has been a gradual yet transformative journey. In the early 20th century, the introduction of radio broadcasts marked the first major shift, allowing fans to experience live game coverage from remote locations. This was followed by the advent of television in the 1950s and 1960s, which revolutionized how audiences consumed sports, bringing real-time visuals into living rooms and vastly expanding fan reach (The impact of radio, n.d.).

As broadcasting technology advanced, innovations such as instant replay, slow motion, and high-definition coverage enhanced the viewing experience and increased the global accessibility of sporting events (Wizevich, 2024). By the late 20th century, sports venues began incorporating basic digital tools—such as electronic scoreboards and rudimentary performance tracking systems—laying the groundwork for deeper integration of technology into the fabric of sports operations.

The 21st-century digital revolution dramatically accelerated this evolution. The proliferation of the internet, social media, and cloud-based platforms redefined fan engagement, while data analytics began playing a central role in athlete performance, coaching strategies, and front-office decision-making. Emerging technologies such as wearable devices, AI, and virtual reality (VR) have since reimagined the sports experience—enabling real-time performance monitoring, immersive fan interaction, and personalized content delivery (Butt, 2024). This fusion of digital innovation with traditional sports practices has created new revenue streams, enhanced global fan engagement, and elevated operational efficiency across the industry. From player scouting to ticket sales, digital tools now underpin nearly every aspect of sport management and consumption.

Today, digital transformation is an essential component for sports organizations seeking to stay competitive, relevant, and sustainable in an increasingly digital and data-driven world. As the pace of technological advancement continues to accelerate, the sports industry stands at the forefront of a broader movement toward intelligent, connected, and fan-centric experiences.

Topical Exploration and Key Concepts

Media Transformation and Digital Fan Experience

The expansion of sports broadcasting beyond traditional television to Over-The-Top (OTT) streaming services has given fans greater flexibility to watch their favorite games on demand (Streaming the game, n.d.). Services like ESPN+, Amazon Prime Video, DAZN, and regional sports networks such as SuperSport in Africa or Hotstar in India have adopted direct-to-consumer (DTC) streaming models, offering live and on-demand sports content from full-match plays to bite-sized highlights. This shift has allowed fans to access games on smartphones, tablets, and smart TVs, eliminating geographical restrictions and enabling global viewership. This trend also results in the cord-cutting trend in favor of digital streaming platforms (Streaming the game, n.d.).

In addition to serving mainstream audiences, OTT platforms provide opportunities for niche sports leagues and emerging sports to reach audiences without the financial barriers associated with traditional television broadcasting (Dear, 2024). This democratization of exposure promotes greater diversity and globalization of the sports ecosystem. Furthermore, OTT streaming also empowers sports organizations to collect detailed viewership data, providing insights into audience preferences, enabling refined content strategies, and enhancing advertising effectiveness. The inclusion of interactive features, such as live polls, real-time chat, and multiangle viewing options, further boosts fan engagement and creates a more immersive experience (Dear, 2024). However, despite improved digital accessibility, challenges persist—including subscription fatigue, content piracy, and technical issues such as network latency, buffering, and platform crashes—which sports organizations must continue to address (Calleja, 2024; Islam, 2024).

Social media platforms like YouTube, Facebook, and Twitter have also introduced live-streaming options, allowing leagues and teams to connect directly with fans. For instance, Major League Baseball (MLB) and the English Premier League (EPL) have partnered with these platforms to stream select games for free, engaging a wider audience. Additionally, social media platforms (e.g., X, Instagram, Facebook, and TikTok) serve as dynamic channels for real-time updates, behind-the-scenes content, and interactive fan experiences (James, 2025). These platforms foster vibrant digital communities that keep fans consistently connected and engaged.

Digital Fan Experience

AI personalizes sports fans' experience by delivering real-time content tailored to user interests. This includes automated highlight reels, game predictions, and customized player statistics. An example is the National Football League's

(NFL) Next Generation Stats, a data-driven platform powered by Radio-Frequency Identification (RFID) chips that tracks player movements, passing metrics, and provides advanced analytics to fans, broadcasters, and teams ("Next Gen Stats", 2024).

Innovations in augmented reality (AR) and VR have also taken the fan experience to new heights, enabling immersive viewing such as VR-based courtside seats or AR-enhanced overlays of player statistics during live broadcasts—for example, the National Basketball Association's (NBA) AR app brings immersive experience by featuring 360-degree AR portals, virtual courtside seats, 3D avatar, and virtual pop-a-shot game (McCaskill, 2023).

Mobile interactive apps also play an important role in fan experiences—offering live scores, ticket purchasing, fantasy integration, and social engagement all in one place. More specifically, the integration of a digital payment system has enhanced the convenience and security of financial transactions in sports. Whether it is purchasing match tickets, merchandise, or food and beverage at a stadium, digital wallets and contactless payment options have streamlined the consumer experience. Many sports arenas now allow fans to preload their digital wallets and make purchases with a tap, reducing wait times and enabling more efficient and smoother in-venue operations. Digital payment systems have also played a crucial role in expanding ecommerce and in-app purchases tied to sports brands and franchises beyond stadiums.

These digital tools not only enhance fan experience—thereby positively influencing fan loyalty and engagement—but also create new revenue streams for sports organizations by leveraging user data to enable targeted advertising, subscription services, and in-app purchases (Power, n.d.).

AI and Big Data Analysis

Performance Enhancement and Injury Prevention

The integration of data analytics and AI is transforming modern sports, particularly in the areas of performance optimization and injury prevention. By leveraging these technologies, teams are moving away from traditional, intuition-based decision-making toward evidence-based strategies that directly impact training regimens, in-game tactics, and long-term athlete development (Raturi, 2023).

In contrast to conventional coaching methods that relied heavily on subjective observation and experience, today's sports organizations utilize real-time, data-driven insights to enhance both individual and team performance. AI-powered platforms such as Second Spectrum, Catapult, and Zebra Technologies leverage high-resolution tracking cameras, wearable sensors, and Global Positioning System (GPS)-enabled devices to capture thousands of data points per second. These include critical metrics such as heart rate, sprint speed, acceleration, deceleration, and even joint angles during movement (Li et al., 2016). For instance, Second Spectrum, an official tracking provider for the NBA and EPL utilizes computer vision and machine learning to provide advanced analytics on shot quality, defensive spacing, and pick-and-roll efficiency.

Beyond performance analysis, predictive analytics plays a crucial role in enhancing strategic decision-making. By modeling various game scenarios and player tendencies, coaches can make more informed, timely decisions during matches (Lee, 2025). In top professional leagues, such as the NBA, NFL, and EPL, data analytics now underpin everything from player recruitment to tactical planning and season-long performance monitoring (The Show, 2025). Notably, these innovations are no longer confined to elite teams; wearable technology and smartphone applications have brought data-driven training within reach for amateur athletes and youth programs as well.

A particularly transformative application of AI in sports is its role in injury prevention and workload management. Machine learning models trained on extensive datasets—including biomechanical readings, game loads, and historical injury records—can detect early signs of fatigue and improper movement patterns (Kaushik, 2024). In the NFL, for example, Zebra Technologies' RFID tracking system is used to monitor impact loads and cumulative physical stress. The resulting data enable medical staff and coaches to adjust training intensity, helping reduce the risk of concussions, muscle strains, and other injuries (Pollard, 2021). These capabilities not only reduce player downtime but also extend career longevity and lessen the financial burden associated with prolonged injuries.

Closely tied to these predictive systems is the increasing use of wearable technology for continuous physiological monitoring. Devices such as GPS trackers, heart rate monitors, and motion sensors offer real-time feedback on key physical performance indicators, including speed, endurance, agility, and recovery rates (Parker, 2024). Among these tools, Catapult has become widely used across soccer, rugby, and American football. Its workload analysis features allow coaching staff to monitor player fatigue levels throughout training and competition, enabling data-informed adjustments to prevent overtraining. Additionally, these wearables provide valuable insights into hydration status, oxygen uptake, and muscular fatigue—critical information for optimizing recovery protocols and maintaining peak performance (Catapult, n.d.).

Taken together, the integration of AI and big data analytics is ushering in a paradigm shift in professional sports—from reactive responses to proactive performance management. These technologies offer sports organizations a powerful competitive advantage by improving athlete health, reducing injury risks, enhancing tactical decisions, and optimizing team performance both on and off the field.

Smart Coaching Tools

AI-assisted coaching has shifted the way athletes train by automating performance assessments and offering personalized training plans (Stengel, 2024). Smart coaching tools such as HomeCourt (for basketball), AlphaPlay (for baseball), and SwingVision (for tennis) provide real-time feedback on an athlete's technique, consistency, and efficiency. For instance, tennis players can analyze their serve speed, spin rate, and shot placement, receiving tailored recommendations for improvement (Swing Vision, n.d.). Similarly, soccer teams use AI-powered tactical boards to simulate game strategies and enhance player decision-making in real-time scenarios.

While coaches have long used analytics to inform their strategies, AI significantly expands the depth and breadth of available data and delivers clear, actionable insights in real time (Stengel, 2024). This allows coaches to make more informed decisions while streamlining aspects of performance monitoring and tactical planning.

Concerns remain that AI tools may diminish the role of a coach's creativity and intuition; However, AI is best viewed as a valuable partner that supports rather than replaces, coaching by enhancing strategic planning and decision-making processes (Stengel, 2024). Importantly, AI-driven coaching platforms are making expert-level training more accessible and affordable, particularly in youth and amateur sports. Athletes no longer need to rely solely on in-person instruction; instead, they can use mobile applications to track their progress, benchmark their performance against elite athletes, and receive personalized, data-informed feedback. As AI technologies continue to evolve, the future of coaching is likely to include AR simulations, virtual training environments, and advanced biomechanical analysis, ensuring that athletes optimize their development at every stage.

Fueled by these innovations, the global sports AI market is projected to grow exponentially, reaching nearly $21 billion by 2029 (Mordor Intelligence, n.d.). This rapid growth reflects not only technological advancement but also the increasing demand for scalable, high-performance solutions across all levels of sport.

Digital Transformation in Officiating

One of the most significant areas where digital transformation has reshaped traditional sports is in officiating. Technological advancements are increasingly being deployed to enhance the accuracy of decisions, reduce human error, and uphold the integrity of competition across various sports disciplines. These innovations exemplify how digital tools are modernizing long-established officiating systems and redefining the role of referees and umpires.

Hawk-Eye Technology

Hawk-Eye is a cornerstone of this transformation. Used extensively in tennis, cricket, and soccer, it employs multiple high-speed cameras and advanced image processing to track the trajectory of the ball in real time. This system provides precise visualizations for in/out calls and goal-line decisions, significantly reducing the margin of human error (Wang, 2024).

Video Assistant Referee (VAR)

The introduction of the Video Assistant Referee (VAR) in elite soccer competitions such as the FIFA World Cup and UEFA Champions League represents another leap in officiating. VAR allows officials to review contentious calls—such as offsides, fouls, and penalties—using slow-motion and multiangle video playback, thereby increasing decision accuracy and transparency (Johnson, 2023).

Automated Ball-Strike (ABS) Systems in Baseball

In MLB, the shift toward digital officiating is seen through the deployment of robot umpires, or Automated Ball-Strike (ABS) systems. These AI-powered tools use optical tracking and machine learning to call balls and strikes with high precision, helping reduce subjective judgment and disputes between players and umpires (Blum, 2024).

Goal-Line Technology

Another major digital innovation is goal-line technology, which is used in soccer and other goal-based sports. This system incorporates sensors and high-speed cameras to determine definitively whether the ball has fully crossed the goal line. The technology provides referees with instant notifications, thereby eliminating the controversy of so-called "ghost goals" (InsideFIFA, 2022).

Instant Replay Systems

Instant replay has become an integral part of digital officiating across multiple sports. These systems allow officials to review critical moments in real time, ensuring that high-stakes decisions are correct. While enhancing fairness, replay technology also introduces new challenges around game flow and the balance between human judgment and technological input (Bell, 2025; Dunaway, 2024). Finding the right balance between technological assistance and human judgment remains an evolving challenge within the broader framework of digital transformation in sports.

Smart Venues: Enhancing Fan Experience through Digital Transformation

The digital transformation of traditional sports has extended beyond gameplay and athlete performance to encompass the entire fan experience, particularly through the development of smart venues. Sports stadiums and arenas around the world are increasingly adopting digital solutions to enhance every stage of the fan journey, from arrival to post-game engagement. These innovations are redefining how fans interact with sports events by improving efficiency, personalization, and convenience across multiple customer touchpoints (Perenzo, n.d.).

To attract and retain spectators in an era of abundant digital entertainment options, sports venues have undergone significant technological upgrades. Digital platforms play a pivotal role in maximizing fan engagement both inside and outside the venue. For example, innovations in egress and ingress management have streamlined crowd flow, allowing fans to enter and exit venues more efficiently. Real-time updates on parking availability, traffic conditions, and public transportation delays are now standard features, enabling smoother arrival and departure experiences (Rankin & Emmett, 2024).

Inside the stadium, digital transformation continues through features that improve wayfinding and in-venue navigation. Mobile apps and smart signage provide directions to seating areas through the least congested entrances, and fans can access real-time information on restroom wait times, restaurant queues, and the availability of merchandise (Rankin & Emmett, 2024). These capabilities enhance fan comfort and reduce frustration, contributing to a more enjoyable event experience.

Moreover, sports organizations are increasingly leveraging big data and AI-driven analytics to better understand consumer behavior and preferences. By analyzing fan interactions, purchase history, and app usage, venue operators can offer personalized, value-added services that cater to individual needs—ranging from targeted promotions to customized food and beverage recommendations (GSIC Report, 2019). This level of personalization strengthens fan loyalty and increases per-capita spending during events.

From a security and operations standpoint, digital platforms also provide essential tools for real-time issue detection and resolution. Fans can report problems—such as safety concerns, lost items, or service delays—directly through venue apps, allowing operators to respond quickly and efficiently (Butler, 2025). This improves not only the overall experience but also the operational safety and responsiveness of the facility.

Crucially, smart venues have embraced digital ticketing systems, enabling seamless access validation and reducing the risk of ticket fraud. Paired with cashless payment systems, these features eliminate the need for physical tickets or cash transactions, resulting in shorter lines, faster service, and enhanced hygiene—a feature especially valued in the post-pandemic environment.

Robotic Process Automation (RPA) for Sports Operations

While much of the focus in the digital transformation of traditional sports centers on athlete performance and fan engagement, Robotic Process Automation (RPA) is driving a revolution behind the scenes. By automating repetitive, rules-based tasks, RPA enables organizations to enhance operational efficiency, reduce human error, and reallocate staff resources to more strategic and value-driven activities (McKinsey & Company, 2017).

In the administrative and management layers of sports organizations, RPA is increasingly being deployed to process contracts, manage payroll, handle procurement, and ensure regulatory compliance. For example, routine functions such as player transfer documentation, visa applications for international competitions, and health compliance reporting can be executed with speed and precision using software bots. This automation minimizes delays and reduces the risk of administrative errors that can compromise competitive and legal standing.

RPA also enhances the accuracy and consistency of data management across departments (McKinsey & Company, 2017). In areas such as financial reporting, ticket sales reconciliation, and fan database maintenance, automated workflows ensure that large volumes of data are processed swiftly and correctly. This is particularly important in a sector where transparency and accountability are paramount—whether to meet the expectations of sponsors, comply with league standards, or satisfy audit requirements from governing bodies.

Beyond internal operations, RPA supports the fan-facing side of digital transformation by integrating with customer relationship management (CRM) systems. Bots can manage season ticket renewals, automate loyalty program updates, and even trigger personalized communications based on fan behavior and preferences. These efficiencies contribute to a more seamless, responsive, and engaging fan experience. For example, McLaren Racing and Automation Anywhere has been working with Formula 1 in the team's race operation (Talisman News Desk, 2019).

In the broader context of digital transformation, RPA exemplifies how sports organizations are evolving into data-driven, tech-enabled enterprises. By streamlining complex administrative workflows and enhancing back-office efficiency, RPA enables teams, leagues, and event organizers to shift their focus from routine tasks to strategic priorities such as growth, innovation, and delivering high-quality experiences to stakeholders both on and off the field.

Trends and Current Issues

The digitization of traditional sports is rapidly transforming the sports industry, presenting both significant opportunities and pressing challenges. As digital technologies become increasingly integral to modern life, sports organizations are being pushed to evolve, not only to remain competitive but also to meet the rising expectations of tech-savvy fans.

Lagging Digital Adoption

According to the digital transformation trends in sports report (Cárdoba, 2025), despite widespread recognition of digital transformation, many sports organizations report slow progress in their digital transformation efforts. A survey of 250 sports executives revealed that organizations rate their digital transformation progress at just 4.3 out of 7, with 60% believing the sports industry is lagging in this area (Cárdoba, 2025). The report also found that while technologies like data analytics and AI are widely viewed as essential for long-term success, only 13% of organizations expressed confidence in their data management and analysis capabilities. This suggests a significant digital skills gap within the industry—one that limits the ability to personalize fan experiences, optimize operations, and drive revenue growth. Without targeted investment in talent development and digital literacy, many organizations risk being left behind.

Development of "Phygital" Sports

The rise of "phygital" sports—hybrid experiences that merge physical game-play with digital environments—is redefining traditional sports boundaries. These experiences combine the physical skill and excitement of real-world sports with the immersive and interactive elements of digital technology. Leagues such as the Metaverse Gaming League (MGL) are at the forefront of this movement, offering innovative formats that blend esports-style gameplay with athletic competition. Phygital sports offer new ways to engage younger audiences who are increasingly drawn to digital entertainment and virtual worlds. They also create opportunities for monetization through Non-Fungible Tokens (NFTs), virtual merchandise, and interactive in-game experiences. However, the integration of phygital formats raises questions about competitive integrity, standardization, and athlete welfare, which must be carefully addressed as this trend evolves (Phygital Sport, 2025).

Providing High Quality Fan Experience

Modern fans expect immersive, on-demand, and personalized experiences—both in stadiums and through digital platforms. Technologies such as AR, VR, and AI-driven content recommendations are being used to meet these expectations. Despite the clear benefits, only 23% of organizations report offering high-quality digital products, and just 17% feel they are effectively engaging fans online (Cárdoba, 2025). Bridging this gap is critical to maintaining relevance in an increasingly competitive entertainment landscape.

To successfully navigate these challenges, sports organizations must invest in scalable digital infrastructure, prioritize innovation, and cultivate in-house expertise. The future of traditional sports lies in how well they can integrate technology to enhance performance, operations, and fan relationships—while preserving the essence of what makes sports special.

Ethical Considerations

While digital advancements offer numerous opportunities, they also introduce various challenges (Hannibal & Knight, 2018), such as data privacy concerns, ethical considerations in AI-driven decision-making, and the increasing commercialization of sports at the potential expense of traditional sporting values. As sports organizations continue to embrace technology, sports managers must carefully evaluate both the benefits and challenges when making decisions about integrating these innovations into their operations.

One of the key ethical concerns in sports technology is data privacy and security. The extensive use of biometric data and AI-driven analytics in athlete performance monitoring raises questions about consent, ownership, and the potential misuse of sensitive information. Sports organizations must implement ethical data governance policies to protect athletes and maintain transparency in data usage. Similarly, fan engagement through digital platforms involves the collection of personal data, which must be protected against breaches and unauthorized use.

The rapid rise of AI in sports betting also raises several ethical and regulatory concerns. The accessibility and convenience of AI-powered betting platforms have contributed to a surge in gambling participation, raising the risk of problem gambling and addiction. AI-driven betting algorithms, while enhancing user engagement, can also exploit psychological tendencies, nudging bettors toward higher-risk wagers through personalized incentives and predictive behavioral models. Furthermore, concerns about match-fixing and integrity in sports have grown, as AI can identify exploitable patterns that could be used for illicit activities. Regulatory bodies and sports organizations must navigate these ethical challenges by implementing responsible gambling measures, transparent AI usage policies, and robust oversight mechanisms to ensure that AI-driven sports betting remains fair, ethical, and sustainable.

As traditional sports embrace digital transformation, stakeholders must navigate these ethical considerations to ensure that technology enhances the industry without compromising fairness, privacy, or the spirit of competition.

Practice in Action

The NBA has been at the forefront of digital transformation, leveraging technology to enhance fan engagement and redefine how audiences experience the game. One of its most notable innovations is the introduction of NBA League Pass, a subscription-based streaming service that allows fans around the world to watch live games, access replays, and enjoy exclusive behind-the-scenes content. This platform has significantly broadened the NBA's global audience and made its content more accessible than ever before.

In 2020, the NBA took a major step forward by partnering with Microsoft to create a more personalized fan experience. Through the integration of AI-powered analytics, fans can access interactive stats, real-time highlights, and AR overlays that provide deeper insights into gameplay (Carney, 2023). This collaboration also enhances content delivery by tailoring it to individual fan preferences.

The NBA's digital savvy mind was also shown during the COVID-19 pandemic. With the restricted in-person attendance, the NBA pioneered the use of virtual fans. Utilizing Microsoft Teams and high-definition digital displays, the league recreated a live-game atmosphere by allowing fans to appear virtually in the arena (Powell, 2010). This innovative approach helped maintain audience engagement during a difficult time and showcased the league's adaptability.

Showcased by these two examples, these forward-thinking initiatives have allowed the NBA to stay ahead of the curve, strengthen fan loyalty, and continuously evolve within an increasingly digital sports ecosystem.

Summary

The digital transformation of traditional sports has revolutionized every aspect of the industry, from athlete performance tracking and fan engagement to business models and governance. Advances in AI, data analytics, blockchain, and immersive technologies have created new opportunities for sports organizations while also presenting challenges related to ethics, privacy, and regulation. As digital trends continue to evolve, sports organizations must balance innovation with maintaining the integrity and values of traditional sports.

Looking ahead, emerging technologies such as the Metaverse, AI-enhanced coaching, and further developments in blockchain are expected to drive the next phase of digital transformation in sports. By adopting a forward-thinking approach, sports organizations can harness technology to create enhanced experiences for players, coaches, and fans while navigating the complexities of the digital age.

Key Resources

- **Giorgio, P. & Westcott, K. (n.d.).** *A whole new ball game: Navigating digital transformation in the sports industry*. Deloitte. https://www.deloitte.com/global/en/Industries/tmt/perspectives/gx-digital-transformation-and-future-changes-in-sports-industry.html
- **Global Sports Innovation Center (GSIC):** https://sport-gsic.com/
- **SportTechie:** https://www.sporttechie.com/
- **NBA Digital Application:** https://www.nba.com/app

References

Bell, J. (2025, February 4). The NFL's tricky technology trajectory means officiating controversies are here to stay. *USA Today*. https://www.usatoday.com/story/sports/nfl/columnist/bell/2025/02/04/nfl-officiating-controversies-technology-advancements/78189678007/

Blum, R. (2024, November 20). MLB will test robot umpire at 13 spring training ballparks hosting 19 teams. *AP News*. https://apnews.com/article/mlb-robot-umpires-experiment-manfred-ecba44e95b37e6f21824cbb58fe69707

Butler, P. (2025, January 29). Game-day defense: Next-gen threat protection for sporting events. *Security*. https://www.securitymagazine.com/articles/101336-game-day-defense-next-gen-threat-protection-for-sporting-events

Butt, A. (2024, October 20). The impact of technology on modern sports: From wearables to AI. *Medium*. https://medium.com/@buttabid728/the-impact-of-technology-on-modern-sports-from-wearables-to-ai-36284de08d5f

Calleja, P. (2024, December 16). Piracy and cybersecurity: Protecting sports broadcast in the cloud era. *Newscaststudio*. https://www.newscaststudio.com/2024/12/16/cybersecurity-piracy-sports-broadcasting-cloud/

Cárdoba, I. (2025). Digital transformation trends in sports report. *GSIC*. https://sport-gsic.com/digital-transformation-trends-in-sports-report-2025-in-collaboration-with-sportian/?_thumbnail_id=47986

Carney, B. (2023, October 25). How the NBA uses generative AI to create personalized content for fans. *Hyscaler*. https://hyscaler.com/insights/nba-uses-generative-ai-for-content/

Catapult. (n.d.). Athletic monitoring system. https://www.catapult.com/solutions/athlete-monitoring

Cespedes-Cubides, A. S., & Jradi, M. (2024). A review of building digital twins to improve energy efficiency in the building operational state. *Energy Informatics*, *7(*11). https://doi.org/10.1186/s42162-024-00313-7

Chen, B., Wan, J., Shu, L., Li, P., Mukherjee, M., & Yin, B. (2018). Smart factory of Industry 4.0: Key technologies, application case, and challenges. *IEEE Access*, *6*, 6505–6519, https://doi.org/10.1109/ACCESS.2017.2783682

Dašić, G. (2023). *Digital Transformation in the Sports Industry*. Sporticopedia.

Dear, D. (2024, July 29). Digital Gold: Turning Olympic moments into lasting legacies for niche sports. *StreamAMG*. https://streamamg.com/news/digital-gold-turning-olympic-moments-into-lasting-legacies-for-niche-sports/

Dunaway, I. (2024, February 27). The debate on instant reply in sports. *The Daily Utah Chronicle*. https://dailyutahchronicle.com/2024/02/27/instant-replay-in-sports/

Frevel, N., Beiderbeck, D., & Schmidt, S. L. (2022). The impact of technology on sports-A prospective study. *Technological Forecasting and Social Change*, *182*. https://doi.org/10.1016/j.techfore.2022.121838

Global Sports Innovation Center. (2019). Digital transformation of sports entities by 2025: What will it look like? https://sport-gsic.com/digital-transformation-of-sports-entities-by-2025/

Hannibal, M., & Knight, G. (2018). Additive manufacturing and the global factory: Disruptive technologies and the location of international business. *International Business Review*, *27*(6), 1116–1127, https://doi.org/10.1016/j.ibusrev.2018.04.003

InsideFIFA. (2022, November 8). Goal-line technology. https://inside.fifa.com/innovation/world-cup-2022/goal-line-technology

Islam, A. (2024, August 30). US fans still think traditional TV is more reliable for live sports than streaming. *SportsPro*. https://www.sportspro.com/news/sports-fans-network-tv-streaming-study-survey-prime-video-peacock-youtube-netflix/

James, B., (2025, February 4). Sports in the age of social media. *The Uproar*. https://nashuproar.org/73487/sports/sports-in-the-age-of-social-media/

Johnson, D. (2023, October 18). How VAR could have changed soccer's most controversial moments. *ESPN*. https://www.espn.com/soccer/story/_/id/38290857/how-var-changed-soccer-most-controversial-moments

Kaushik, V. (2024, April 4). Predictive analytics and injury prevention: Leveraging AI to identify potential risks in fitness apps. *Forbes*. https://www.forbes.com/councils/forbestechcouncil/2024/04/04/predictive-analytics-and-injury-prevention-leveraging-ai-to-identify-potential-risks-in-fitness-apps/

Lee. S. (2025, March 27). Top 7 predictive analysis trends revolutionizing sports analytics. *Number Analytics*. https://www.numberanalytics.com/blog/trends-predictive-sports-analytics

Li, R. T., Kling, S. R., Salata, M. J., Cupp, S. A., Sheehan, J., & Voos, J. E., (2016). Wearable performance devices in sports medicine. *Sports Health*, *8*(1), 74–78. https://doi.org/10.1177/1941738115616917

McCaskill, S. (2023, February 20). NBA's AR tech lets fans add themselves into live games. *SportsPro*. https://www.sportspro.com/news/nba-ar-live-stream-avatar-immersion-app-personalisation-all-star/

McKinsey & Company. (2017, March 1). The value of robotic process automation. https://www.mckinsey.com/industries/financial-services/our-insights/the-value-of-robotic-process-automation

Mordor Intelligence. (n.d.). AI in sports market size & share analysis-growth trends & forecasts (2025–2030). https://www.mordorintelligence.com/industry-reports/artificial-intelligence-market-in-sports

Next Gen Stats. (2024, October 9). *Wikipedia*. https://en.wikipedia.org/wiki/Next_Gen_Stats

Parker, S. (2024, October 9). Wearable technology options for athletes continue to expand. *NFHS*. https://www.nfhs.org/articles/wearable-technology-options-for-athletes-continue-to-expand/

Parviainen, P., Tihinen, M., Kääriäinen, J., & Teppola, S. (2017). Tackling the digitalization challenge: How to benefit from digitalization in practice. *International Journal of Information Systems and Project Management*, *5*(1), 63–77. https://doi.org/10.12821/ijispm050104

Perenzo, P. (n.d.). How artificial intelligence is transforming the sports industry. *Imaginovation*. https://imaginovation.net/blog/ai-in-sports-industry/

Phygital Sport. (2025, March 13). *In Wikipedia*. https://en.wikipedia.org/wiki/Phygital_sport?utm_source=chatgpt.com

Pollard, J. (2021, November 24). Why NFL players are benefiting from RFID technology. *ZEBRA*. https://www.zebra.com/us/en/blog/posts/2021/why-nfl-players-are-benefitting-from-rfid-technology.html

Powell, S. (August 10, 2010). How virtual fans found their seats at NBA season restart. *NBA.Com*. https://www.nba.com/news/virtual-fans-help-restart-atmosphere

Power, R. (n.d.). Five simple ways to increase revenue in live sports. *Dolby Optiview*. https://optiview.dolby.com/resources/blog/streaming/five-simple-ways-to-increase-revenue-in-live-sports/

Queiroz, M. M., Fosso Wamba, S., Machado, M. C., & Telles, R. (2020). Smart production systems drivers for business process management improvement: An integrative framework. *Business Process Management Journal*, *26*(5), 1075–1092. https://doi.org/10.1108/BPMJ-03-2019-0134

Rankin, J., & Emmett, J. (2024, July 24). How technology is redefining the fan experience at sports and music venues. *Gensler*. https://www.gensler.com/blog/immersive-technology-is-redefining-the-sports-fan-experience

Raturi, G. (2023, March 30). Artificial intelligence in sports training. *Medium*. https://medium.com/geekculture/artificial-intelligence-in-sports-training-2d7736734819
Stengel, G. (2024, May 15). Sports AI can be a game-changing partner for coaches. *Forbes*. https://www.forbes.com/sites/geristengel/2024/05/15/sports-ai-can-be-a-game-changing-partner-for-coaches/
Streaming the game. (n.d.). *PWC*. https://www.pwc.com/us/en/industries/tmt/library/sports-streaming-platforms.html
Swing Vision. (n.d.). *ITA Tennis*. https://wearecollegetennis.com/ita/partners/swingvision/
Talisman News Desk. (2019, December 30). McLaren racing and automation anywhere announce Formula 1 partnership. *Talisman*. https://www.thetalismanagency.com/playbook/mclaren-racing-and-automation-anywhere-announce-formula-1-partnership
The impact of radio on baseball in the early 20th century. (n.d.). Historic baseball. https://historicbaseball.com/the-impact-of-radio-on-baseball-in-the-early-20th-century/
The Show. (2025, January 24). The evolution of data-driven decision-making in sports. https://theshowvdgsports.com/the-evolution-of-data-driven-decision-making-in-sports/
Vial, G. (2019). Understanding digital transformation: A review and a research agenda. *The Journal of Strategic Information Systems*. *28*(2), 118–144.
Wang, L. (2024, March 6). Hawk-eye: A look into the future. *The Bishop's Tower*. https://thebishopstower.com/6617/sports/hawk-eye-a-look-into-the-future/
Wizevich, E. (2024, December 7). When instant reply debuted during the broadcast of a college football game in 1963, it revolutionized the way we watch sports. *Smithsonian Magazine*. https://www.smithsonianmag.com/smart-news/when-instant-replay-debuted-during-the-broadcast-of-a-college-football-game-in-1963-it-revolutionized-the-way-we-watch-sports-180985516/?utm_source=chatgpt.com

9

SPORT TECH STARTUPS AND INNOVATION

Matthew A. Masucci and jay johnson

Introduction

The intersection of sport and technology is experiencing a transformative shift, driven in large part by the rapid emergence of startups, defined here as early-stage, innovation-driven companies created to scale quickly and disrupt existing markets (Butler, 2000; Ratten, 2020a). These ventures are reimagining how athletes train, how fans interact, and how organizations manage performance through digital platforms, wearable devices, biometric monitoring, and AI-driven tools (Chopra et al., 2021; Ferreira et al., 2020). Situated within the broader sport-business and sport-technology ecosystem, these startups act as agile disruptors, often outpacing larger companies in their ability to innovate and adapt (González-Serrano, Añó Sanz, and González-García, 2020).

The global sport technology sector is expanding rapidly. In 2024 alone, US sport-tech startups secured 119 deals totaling $880 million in venture funding, demonstrating investors' confidence in the market's potential (Business Insider, 2025). Moreover, in 2024, there were around 62 million fitness tracker (a wearable device that monitors your physical activities and provides data on your fitness progress) users in the United States, most of whom were smartwatch owners. In the near future, the number of fitness tracker users is forecast to steadily increase, jumping to over 92 million by 2029 (Statista, 2024). Key areas of growth include artificial intelligence (AI), machine learning (ML), the Internet of Things (IoT), and immersive technologies such as virtual reality (VR) and augmented reality (AR) (Flynn, 2024; MasterBorn, 2023). Startups are leveraging these innovations to offer hyper-personalized, data-driven services that appeal to athletes, coaches, fans, and recreational users alike.

In the context of sport technology, advanced algorithms refer to data-driven computational techniques that utilize AI and ML to process complex and multimodal data (such as health sensor signals, video footage, and audio streams) in order to assess, predict, and enhance athletic performance, team

DOI: 10.4324/9781003588405-11

strategy, and game dynamics. AI is broadly defined as the field of computer science concerned with building systems capable of performing tasks that typically require human intelligence, including perception, reasoning, and decision-making. In a similar manner, this book considers ML as a subfield of AI that focuses on the development of algorithms that allow systems to automatically learn from data and improve their performance on a task over time without being explicitly programmed (Li & Huang, 2024). These algorithms in sport analytics span a range of methodologies, including statistical learning, deep learning, and reinforcement learning, and are increasingly applied in real-time environments to generate insights from raw performance data and provide actionable feedback to coaches, athletes, and support staff (Ghosh et al., 2023).

Relatedly, the IoT represents a transformative paradigm in modern technology, characterized by a vast network of interconnected devices that collect, exchange, and act upon data to enhance efficiency and decision-making across various sectors. Recent academic literature underscores IoT's integration with emerging technologies such as AI, edge computing, and blockchain, facilitating advancements in smart cities, healthcare, and industrial automation (Oliveira et al., 2024). However, challenges persist, notably in ensuring data security, privacy, and the development of scalable architectures to manage the increasing complexity and the volume of IoT devices (Choudhary, 2024). Yet, alongside these opportunities come ethical and regulatory challenges, the increased use of biometric and behavioral data introduces pressing questions about privacy, surveillance, and informed consent. AI systems in sport also risk reinforcing bias and lack transparency in decision-making processes that can affect athlete evaluation or fan experience (SVG Europe, 2023). These concerns are magnified in a regulatory landscape that is only beginning to catch up with the speed of innovation.

This chapter addresses how sport-tech startups are balancing technological advancement with ethical responsibility in an evolving market. This topic is not only urgent and consequential for founders and investors making strategic decisions but also for governing bodies seeking regulatory clarity, consumers expecting transparency, and scholars examining the future of digital sport (Ratten & Thompson, 2020). In addition, this chapter explores how sport-tech startups are transforming the sport and fitness industries through technological innovation, while also grappling with complex ethical and regulatory challenges. It delves into the defining features of sport startup ventures, such as their distinctive business models, innovative approaches, and the ways they disrupt traditional modes of physical activity. Alongside this, the chapter examines the ethical dilemmas and regulatory hurdles that often accompany the rapid deployment of emerging technologies in sport. Through case studies and analysis of influential companies, the chapter provides insight into how these startups are not only changing how individuals engage with sport and fitness but also shaping the broader future of physical activity in society. By tracing historical developments, identifying core concepts, and highlighting current trends, the chapter offers a comprehensive perspective on the evolving sport-tech startup landscape and its long-term implications.

Background

Sport-Tech Origins and Early Technological Milestones

The integration of technology into sport began long before the advent of modern startups, with foundational innovations in the 20th century that emphasized accuracy and efficiency in both training and competition, and some early advancements included electronic timing systems introduced at the 1912 Stockholm Olympics and the development of photo-finish cameras by the 1948 London Games (Toohey & Veal, 2007). These innovations, among others, established a baseline of precision that influenced future technologies (Crawford et al., 2015). Another pivotal moment was the introduction of broadcast replays in 1963 by Columbia Broadcasting System (CBS), which revolutionized sports viewing by allowing audiences and officials to rewatch crucial moments, and these developments laid the groundwork for integrating more complex technologies such as video assistant referees (VAR) and performance analytics tools (McChesney, 2004). The commercial application of wearable technology began in 1977 with Polar Electro's creation of the first wireless ECG heart rate monitor, designed to assist the Finnish National Cross-Country Ski team. This innovation marked the beginning of personalized physiological monitoring in sport and catalyzed further research and product development in the domain of biometric wearables (Laukkanen & Virtanen, 1998; Motti, 2020).

The Rise of Sport-Tech Startups

The 2000s saw a significant acceleration in the sport-tech sector, driven by the convergence of broadband internet, smartphone penetration, and cloud computing, where these technologies lowered the barrier to entry for startups, allowing new players to leverage data, community features, and cloud-based analytics to disrupt traditional sport paradigms (Ratten, 2024). Companies like Catapult Sports and Strava exemplify this transformation with Catapult, founded in 2006, introducing GPS and wearable tracking systems to monitor athletes' biomechanics and workload and offering insights that were previously accessible only to elite teams (Aughey, 2011). Similarly, Strava, launched in 2009, utilized GPS-enabled smartphones to create a social fitness platform that attracted a global user base and democratized access to performance analytics (Maguire, 2015).

The maturity of sport-tech as a sector has been aided by dedicated accelerators and venture capital funds that provide startups with funding, mentorship, and market access, seeing programs like Stadia Ventures, HYPE Sports Innovation, and Techstars Sports have significantly contributed to the proliferation of sport-tech ventures globally (Ratten, 2020b). Moreover, the convergence of AI, wearables, and immersive technologies like AR and VR is shaping the next frontier with AI-driven platforms such as Playermaker and Whoop that use predictive analytics to enhance training regimens and reduce injury risk, while VR-based tools offer immersive training environments for skill acquisition and rehabilitation (Xiao et al., 2023).

The Pandemic Catalyst and Global Innovation Hubs

During the COVID-19 pandemic, digital adoption in sport saw a massive acceleration. Lockdowns and social distancing measures catalyzed a widespread shift toward at-home fitness solutions and direct-to-consumer models (Wang, 2021), resulting in exponential growth for companies like Peloton and Strava as consumers embraced virtual communities and digital coaching experiences (Glebova et al., 2024). The pandemic underscored the importance of adaptability and remote engagement in the sport-tech landscape.

Major innovation clusters have emerged in cities such as London, Paris, Tel Aviv, and San Francisco, catalyzed by robust ecosystems of startups, academic institutions, corporate partners, and sports governing bodies (Ratten, 2021a). These hubs are instrumental in nurturing the convergence of digital health, athletic performance, fan engagement, and media technologies. This expansion is further propelled by increasing consumer demand for health- and data-driven solutions, particularly among digitally native and fitness-oriented populations (Deloitte, 2021).

Ethical and Regulatory Challenges in Sport-Tech

One of the most pressing challenges facing sport-tech startups today concerns data privacy and the ethical use of biometric and performance data; the proliferation of wearable devices and tracking software has enabled real-time monitoring of athletes' physiological, psychological, and biomechanical states (Ratten, 2020c). While these innovations offer substantial benefits for performance optimization and injury prevention (Berg, 2022; Seshadri et al., 2017, 2019), they also raise critical concerns regarding the ownership, consent, and secondary use of athlete data.

The European Union's General Data Protection Regulation (GDPR) has introduced significant legal obligations for companies processing personal data, including the biometric data commonly collected by sport-tech firms (Mann et al., 2020). Failure to comply with these regulations not only jeopardizes user trust but also exposes organizations to substantial legal and financial penalties, ensuring privacy-by-design frameworks and transparent consent protocols have become a key differentiator in this competitive landscape (Miller et al., 2024).

Beyond privacy, the lack of unified regulatory frameworks for athlete monitoring and sport-tech deployment poses another major risk as there is currently only a patchwork of policies across jurisdictions regarding what constitutes ethical data collection and permissible use in both professional and amateur sport (Jessop & Baker, 2019). This regulatory fragmentation complicates international scalability for startups and creates uncertainty for investors and sports organizations alike (Ratten, 2019).

Moreover, concerns have been raised about the potential for surveillance-based coaching to erode athlete autonomy and create exploitative dynamics, especially in elite youth development contexts (Gard et al., 2020). As such, there is an urgent need for sport governing bodies and regulatory agencies to collaborate in creating globally recognized standards for technology use in sport, balancing innovation with fairness, safety, and athlete welfare.

Business Models, Market Saturation, and Strategic Futures

The monetization strategies of sport-tech startups are also under increasing scrutiny, with subscription models dominating the market, particularly among fitness apps, digital coaching platforms, and connected hardware providers. However, user retention remains a key challenge, as consumers often exhibit low switching costs and subscription fatigue and startups must continuously innovate their value proposition, often combining personalized feedback, gamification, and community engagement to sustain customer loyalty (Napoli, 2022). Additionally, as the market matures, it is becoming increasingly saturated with offerings that often overlap in functionality, and this saturation can lead to market fragmentation and pressure to consolidate, especially among venture-backed firms facing tightening funding conditions amid broader macroeconomic volatility (Lokey, 2024).

The future of sport-tech startups will likely depend on their ability to integrate ethical considerations into their core business models, including not only compliance with evolving legal standards but also proactive engagement with athletes, coaches, and fans regarding data practices, algorithmic decision-making, and equitable access seeing strategic alignment with public health priorities. Imperatives such as inclusive fitness, mental well-being, and injury prevention may also unlock new partnerships and funding streams (Glebova, Lee, & Kim, 2024).

To navigate this complex terrain, startups must invest in regulatory expertise, ethical design, and interdisciplinary collaboration. Doing so will not only mitigate risk but also build trust and long-term resilience in an industry poised at the intersection of sport, health, and technology (Glebova, Lee, & Kim, 2024).

Topical Exploration and Key Concepts

The Sport-Tech Innovation Landscape

Sport-tech startups are situated within a rapidly evolving innovation ecosystem that spans the domains of sport, technology, health, and business (Martín-Peña et al. 2024; Glebova & Desbordes, 2021). These startups often emerge at the intersection of performance science, consumer technology, and fan engagement, where new ideas can be prototyped, tested, and scaled quickly (Ratten, 2021a). A sport-tech ecosystem typically includes venture capitalists, angel investors, incubators, accelerators, corporate innovation units, and public sector stakeholders (Sethi & Talvelkar, 2024; Campo-Beamud & Rodrigo-Carranza, 2024). Sport technology startups often rely on a diverse ecosystem of investors and support organizations to fuel their growth, including venture capitalists, angel investors, incubators, accelerators, corporate innovation units, and public sector stakeholders.

Venture capitalists are professional investors who manage independent, dedicated pools of capital that focus on equity investments in privately held, high-growth companies (Gompers & Lerner, 2001). Some prominent venture capitalists that have invested in sports technology are Sequoia Capital, which backed WHOOP, Andreessen Horowitz, which invested in Zwift, and Forerunner Ventures, which supported Mirror. Angel investors such as LeBron James (Tonal), Ashton Kutcher (SmartSpot), and Serena Williams (Mahmee) are wealthy individuals who invest their own personal funds into startups at early stages (often preceding formal venture capital) in exchange for equity and mentorship (Wetzel, 1983). Incubators (such as Hype Sports Innovation, MIT Sports Lab, and leAD Sports Accelerator) are specialized programs or organizations designed to support early-stage startups by offering resources such as physical workspaces, mentorship, access to financing, and networking opportunities (Grimaldi & Grandi, 2005).

Accelerators are fixed-term, cohort-based programs that provide startups with mentorship and educational components to rapidly scale their business, typically culminating in a public pitch event or demo day (Cohen & Hochberg, 2014). Examples of sport technology accelerators that have been on the leading edge of the industry include Techstars Sports, Nike, and Valiant Labs. Corporate innovation units (e.g., corporate incubators), meanwhile, are dedicated divisions within established companies that foster new businesses by providing internal startups with physical resources and support to drive innovation in the parent firm (Weiblen & Chesbrough, 2015). Some prominent corporate innovation units are Adidas Future Lab, Under Armour Connected Fitness, and Peloton Labs.

Finally, public sector stakeholders like the UK Sport Institute, US Olympic Committee, and Sport Canada are government agencies and other public institutions that facilitate startup development by offering supportive policies, funding initiatives, and infrastructure that encourage innovation in emerging sectors like sport technology (Mazzucato, 2013; Ratten, 2023).

Startup Investment Ecosystem

A major driver of this funding activity is the belief that technology-enabled products and services can radically transform sport training, injury prevention, data analytics, and audience experiences (Huang & Yongquan, 2025; Sethi & Talvelkar, 2024). Startups in the sport technology space often benefit from nontraditional funding avenues, seeing corporate venture capital from firms such as Adidas, Nike, and Under Armour has become commonplace, with these firms investing in up-and-coming ventures to stay on the cutting edge of sport innovation (Baena-Arroyo et al., 2020). Additionally, fan-driven crowdfunding has emerged as a unique feature of this sector, with consumers funding new wearable devices, mobile fitness apps, or smart training equipment—essentially validating demand before a product hits the market. Crowdfunding platforms such as Kickstarter and Indiegogo have also hosted campaigns for sport-tech innovations ranging from connected shoes to AI-powered coaching systems (Gazzola et al., 2024; Troise et al., 2022).

The startup ecosystem also benefits from sport-specific accelerators such as Stadia Ventures or leAD Sports, which provide funding, mentorship, and networking tailored to the unique needs of sport-tech founders (Ratten, 2021b). This growing support structure has helped create a virtuous cycle of innovation and investment, positioning sport-tech startups as vital contributors to the broader digital transformation of sport.

Emerging Business Models and Revenue Strategies

Sport-tech startups frequently differentiate themselves through innovative, service-oriented business models that go beyond traditional product sales, in contrast to conventional sporting goods companies (Huertas González-Serrano et al., 2019; Ratten & Jones, 2020). Typically generating revenue from one-time purchases of equipment, sport-tech ventures increasingly employ models centered on recurring revenue and customer lifetime value, including subscription models, data-as-a-service offerings, and digital ecosystems that foster ongoing user engagement (Ratten, 2024).

One emblematic example is Peloton, which has revolutionized home fitness by integrating connected hardware (e.g., smart bikes and treadmills) with a robust content subscription platform (Hutton, 2021). While Peloton generates revenue from equipment sales, its primary growth driver has been its subscription base, which provides access to live and on-demand classes, community features, and gamified performance metrics (Hutton, 2021; Winchester et al., 2021). This approach exemplifies a shift in value proposition—from delivering a product to delivering an ongoing experience.

WHOOP, a startup offering continuous biometric monitoring, followed a similar path. Initially priced as a premium wearable, WHOOP transitioned in 2018 to a subscription-based model, providing the hardware "free" with a $30/month membership that includes data analytics, health insights, and recovery coaching (WHOOP, 2019). This model transforms a discrete product into a long-term service relationship, aligning profitability with user outcomes.

Newer startups are building on this logic with hybrid revenue streams. Zwift, the virtual training platform, not only sells access to immersive cycling and running environments via subscription but also monetizes its user base through in-app purchases, branded events, and even esports tournaments (Westmattelmann et al., 2021). Similarly, Strava began as a freemium model-providing core functionality for free while charging for premium features such as personalized training plans and deep performance analytics but has since pivoted to a membership subscription model once a critical mass of users had signed up (Peckham, 2019).

What unites these models is a focus on building ecosystems around the product where sport startups create digital platforms where user data, social connections, and content intersect, deepening engagement and lowering churn (Baena-Arroyo et al., 2020). For entrepreneurs, this approach offers advantages in customer retention, monetization, and brand loyalty, all of which are critical to startup scalability in highly competitive tech markets. Moreover, in the startup context, a product ecosystem refers to a network built around a core product—typically a digital platform—that includes related services, features, and partner contributions. Sport tech startups often design their platforms to support integrations with third-party tools, data sources, and content, enhancing functionality and user engagement (Hajli, 2025). This ecosystem approach helps startups retain customers, encourage repeated use, and diversify revenue streams. As Martín-Peña et al. (2024) observe, digital platforms that foster ecosystem development enable startups to create greater value and improve their competitive position in rapidly evolving markets.

Biometric Analytics and Wearable Innovation

Biometric analytics, the process of collecting, interpreting, and applying physiological data, serves as a foundational element in the development of many sport-tech startups (Malik et al., 2024). In the sport technology domain, this practice encompasses the analysis of physiological and biomechanical indicators such as heart rate variability (HRV), muscle activation patterns, core body temperature, and respiratory rate (Sinar & Acar, 2024b). These metrics are increasingly captured through wearable sensors and processed via real-time data platforms to deliver nuanced insights that can enhance athletic performance, optimize recovery protocols, and reduce injury risk. For athletes and coaches, the integration of biometric analytics into training regimens allows for individualized feedback, data-driven decision-making, and evidence-based adjustments to workload and technique. As emphasized by Malik et al. (2024), the growing sophistication of wearable technologies and data analytics tools has significantly improved the accuracy and applicability of biometric monitoring in elite and recreational sports settings.

Recent studies highlight these advances. Nithya and Nallavan (2023) developed a wearable system that tracks muscle strain and vital signs to help prevent injuries during training. Similarly, Espinosa et al. (2025) emphasize the growing role of wearable sensors in delivering personalized, data-driven training programs, which have helped to underpin and advance the wearable technology sector (Malik et al., 2024).

Wearable technologies, powered by advances in miniaturized sensors and edge computing, now allow athletes and everyday users to continuously track metrics such as HRV, sleep cycles, blood oxygen levels, and neuromuscular load (Seshadri et al., 2019). This proliferation of data has opened new markets for startups that can turn raw metrics into actionable insights. Startups like Zone7, used by elite teams in soccer, cycling, and American football, have pioneered the application of AI to biometric data streams, enabling predictive analytics that can forecast injury risk and recommend adjustments to training loads to fine-tune performance and prevent burnout (García-Peñalvo, 2021). While at the consumer level, WHOOP, Fitbit, and Oura (all examples of wearable technology) offer similar services to individuals, positioning themselves as personal coaches embedded in wearable form.

Sport technology startups distinguish themselves not only by the precision of their data collection methods but also by the sophistication of the algorithms they employ to process and interpret this information. A growing number of these ventures now incorporate multiple streams of data, including biometric signals such as heart rate and muscle activation, environmental variables such as temperature and altitude, and behavioral indicators such as movement patterns and decision-making tendencies. By synthesizing these diverse inputs, startups are able to construct comprehensive and dynamic profiles of an athlete's physical and cognitive readiness. The resulting data are then translated into accessible visual formats through intuitive digital interfaces, enabling users, athletes, coaches, and performance staff to make informed decisions and set measurable, evidence-based training and performance goals (Huang, 2022).

Academic research supports the efficacy of such approaches, where Pickering and Kiely (2019) argue that individualized training guided by wearable biometrics can improve performance and reduce the risk of overtraining

(Sinar & Acar, 2024a). The growing adoption of such technologies at both professional and amateur levels suggests a new paradigm in sport science—one where startups play a pivotal role in democratizing access to high-quality performance insights (Allouch, 2024).

Gamification and User Engagement

Gamification is the use of game elements such as points, badges, levels, and leaderboards in non-game contexts and represents a core strategy used by sport-tech startups to boost user motivation, enjoyment, and retention (Sardi et al., 2017). In the sport-tech domain, gamification helps transform routine physical activity into an engaging, competitive, and socially rewarding experience, which is particularly crucial for startups seeking to differentiate themselves in a crowded digital fitness market and reduce user churn, which can be as high as 70% in the first three months for fitness apps (Edney et al., 2019).

Startups like Zwift and Strava exemplify this strategy, converting solo indoor cycling into a multiplayer game, immersing users in virtual worlds where they can race, train, and socialize that integrates elements such as avatars, virtual rewards, levels, real-time feedback, and global leaderboards. This converts an otherwise monotonous exercise routine into a competitive, immersive experience, making it more likely that users will return consistently (Westmattelmann et al., 2021). Zwift has also launched structured challenges and esports competitions, such as the UCI Cycling eSport World Championships, creating new pathways for competition and brand loyalty (Westmattelmann et al., 2021). From a business model perspective, gamification increases the perceived value of the subscription service and facilitates upselling of additional features or merchandise.

Strava leverages social gamification through its "segments" feature, which enables users to compete asynchronously with peers and strangers on specific routes, and these digital trophies and rankings turn everyday exercise into a quest for improvement and recognition. These features foster a sense of community and accountability, the two factors known to drive adherence to exercise regimes (Zuckerman & Gal-Oz, 2014). Moreover, data generated from gamified interactions offer startups critical behavioral insights that can power personalized recommendations and targeted marketing strategies (Singh & Khan, 2022). By tracking user engagement patterns, such as task completion, interaction frequency, and reward responses, startups can refine user segmentation and customize experiences in real time. Recent studies show that gamification analytics enhance user profiling, supporting adaptive content delivery and more precise marketing outreach (Hajarian et al., 2019). In sport-tech, where user motivation and engagement are central, these insights help optimize retention and drive monetization through tailored challenges and dynamic messaging (Farhan et al., 2023).

For startups, gamification provides more than just engagement; it can be a key lever for behavior change and value creation, as studies have shown that gamification elements, such as progress tracking and competition, can increase physical activity levels by over 20% (Johnson et al., 2016; Sardi et al., 2017). This is particularly impactful for early-stage companies aiming to prove efficacy and secure user base expansion. Importantly, successful gamification is not just about extrinsic rewards; it also taps into intrinsic motivators like mastery, autonomy, and relatedness—factors highlighted in self-determination theory (Behzadnia et al., 2022).

From an innovation standpoint, sport-tech startups are well-positioned to experiment with adaptive gamification where personalizing game mechanics is based on user preferences and behavior patterns, where ML can dynamically adjust challenges, feedback, or incentives to optimize motivation for different user segments (Johnson et al., 2016). The functionality not only boosts retention but creates defensible IP and unique value propositions—critical elements for startups seeking to attract investment or achieve product-market fit (Sardi et al., 2017).

Ethical and Regulatory Considerations

The proliferation of sport-tech startups and their data-intensive products raises significant ethical and regulatory challenges, particularly around privacy, equity, and governance (Berg & Fischer, 2023). These challenges are heightened by the fact that many startups operate outside traditional regulatory regimes, navigating legal grey areas around biometric data collection, algorithmic decision-making, and virtual competition (Clark, 2021).

One of the most pressing issues is data privacy as wearables and fitness apps routinely collect sensitive health and location data. The entities collecting this information, which are often startups, are not classified as "covered entities" under laws like the Health Insurance Portability and Accountability Act (HIPAA) in the United States (Arnow, 2016). As a result, consumer data protections often depend on corporate privacy policies rather than legal mandates (Sifaoui & Eastin, 2024). The WHOOP–Major League Baseball partnership, for example, had to establish a unique data governance framework to address ownership and consent, underscoring the need for clear standards in sport-tech applications (Seshadri et al., 2019).

Ethical challenges extend beyond privacy. As high-performance technologies become increasingly embedded in elite and recreational sport, access disparities may exacerbate inequality where athletes or teams with the financial resources to adopt cutting-edge biometric devices or AI-driven coaching tools could gain competitive advantages over less-equipped peers, creating a "technology divide" (García-Peñalvo, 2021). For startups, this raises questions about inclusivity and social responsibility. Those that actively design for accessibility, such as offering tiered pricing or open-source versions, can help democratize technology use while gaining broader market appeal (Clark, 2021).

Another ethical concern is "tech doping," wherein users manipulate technological systems for unfair advantage. For example, in the context of Zwift races, users have been caught falsifying weight inputs or using bots to inflate performance, which is akin to digital cheating (Westmattelmann et al., 2021). This problem not only undermines trust in digital competition but also puts pressure on startups to implement robust anti-fraud systems, ranging from biometric authentication to AI-driven anomaly detection.

As the sports technology industry evolves, startups must proactively address ethical and regulatory challenges to ensure sustainable growth. Integrating ethical principles into product development—such as privacy-by-design, fairness auditing, and transparency—can foster user trust and serve as a competitive differentiator (Najjar, 2023; Sampedro, 2023). For instance, implementing privacy-by-design ensures that data protection measures are embedded throughout the system development process, aligning with legal requirements and user expectations (Saltarella et al., 2021). Moreover, conducting fairness audits on algorithms can help identify and mitigate biases, promoting equitable outcomes (Yan & Zhang, 2022).

Transparency in data collection and usage is particularly vital. Clear, user-friendly disclosures about what data is collected, how it is used, and who it is shared with are key to ethical digital innovation (Reischauer et al., 2024). Over time, the establishment of industry-wide standards and voluntary codes of conduct will likely become essential. Startups should actively participate in shaping these frameworks, helping to guide responsible innovation rather than reacting to imposed constraints. This proactive approach not only mitigates legal risks but also demonstrates leadership in ethical sport technology development (Lad & Caldwell, 2009; Bouhaj et al., 2022).

Trends and Current Issues

Digital Innovation and the Transformation of Sport

Sport-tech startups are not merely innovating within existing market boundaries; they are actively reshaping the future of how sport, fitness, and physical activity are defined, practiced, and experienced (Frevel et al., 2023; Hammerschmidt et al., 2024). Through digital transformation, these ventures are lowering barriers to entry, creating new forms of participation, and shifting the cultural meanings of athleticism and health (Glebova et al., 2025). For instance, the integration of big data and social media has altered the way entrepreneurship is embedded in a sports context, leading to increased interactivity and accessibility (Esmer & Dertli, 2024; Ratten, 2020c). Moreover, the convergence of sport, fitness, and lifestyle sectors has given rise to unique environments for deploying entrepreneurial thinking and creative business models, emphasizing the role of cultural and social trends (Jones et al., 2020). As Cossich et al. mention (2023), emerging technologies such as AI, VR, and AR are redefining sport performance analysis and creating immersive training environments. These advancements not only enhance elite competition but also have significant implications for grassroots fitness and societal well-being, as they promote ethical considerations and strategies for safeguarding athlete welfare (Qi et al., 2024). The Global Sports Innovation Center projects that by 2025, digital transformation will be integral to sport entities, influencing aspects from fan engagement to athlete performance (GSIC, 2019).

Expanding Access, Inclusion, and Hybrid Fitness Models

One of the most significant impacts of sport-tech startups is their role in expanding access to physical activity. Companies like Peloton and Zwift have used technology to reconfigure time, space, and community, allowing users to engage in structured fitness without geographic or scheduling constraints (Methlagl et al., 2025; Supriyanto & Liu, 2022). Peloton's streaming platform transformed solitary at-home workouts into socially connected, instructor-led experiences. During the COVID-19 pandemic, its membership soared, rising from 1.4 million in 2019 to over 5 million by 2021, highlighting how sport-tech startups can scale quickly by fulfilling latent demand for accessible, engaging fitness options (Hutton, 2021; Winchester et al., 2021).

This democratization of fitness also reflects a broader shift toward hybrid and asynchronous exercise models. Many users now combine physical and digital routines, for example, gym sessions supplemented by app-based programs or virtual races that complement in-person competitions (García-Peñalvo, 2021). This blended fitness ecosystem opens the door for startups to serve a more diverse set of consumers, including those with disabilities, time constraints, or geographic isolation. Startups that design with inclusion in mind, such as adaptive interfaces, on-demand customization, or language localization, are not only advancing equity but also accessing underserved markets (García-Peñalvo, 2021).

The Phygital Experience and Pandemic Acceleration

Sport-tech startups are also eroding the traditional boundaries between physical and virtual sport. Platforms like Zwift, which hosts competitive esports cycling, and VR-based training companies like Rezzil and Rouvy, illustrate how digital and physical exertion are merging into hybrid athletic experiences (Cossich et al., 2023). The 2020 UCI Cycling eSport World Championship (*Cyclingnews*, 2019) was a landmark event, legitimizing digital racing as a professional sport. Startups are thus not only creating new forms of competition but also generating new revenue streams through virtual spectatorship, licensing, and event sponsorship (Westmattelmann et al., 2021).

This blending of the physical and the digital has given rise to a concept termed *phygital.* Phygital is a blend of "physical" and "digital"—describes experiences that integrate real-world activities with digital technologies to enhance engagement and interaction (Batat, 2022; Mele et al., 2023). In the sport and fitness industries, phygital innovations are transforming how individuals exercise, train, and compete; for instance, Peloton combines physical exercise equipment with live and on-demand digital classes, performance tracking, and social features to foster motivation and community among remote users (Pizzo et al., 2020). Similarly, platforms like Zwift enable users to engage in physical workouts while interacting with others in virtual environments, participating in races, following training plans, and tracking biometric data. (Westmattelmann et al., 2021).

As we mentioned in the key concepts section, the COVID-19 pandemic accelerated the adoption of such hybrid models. With fitness centers closed and events canceled, individuals turned to virtual platforms to stay active, prompting a surge in digital fitness adoption. Research indicates that physical activity levels declined globally during the pandemic, making digital tools essential for sustaining exercise routines (Wilke et al., 2021). Moreover, sports organizations and youth programs adapted by offering online training and wellness programs, supporting continued physical, social, and emotional development during lockdowns (Bates, Greene, & O'Quinn, 2021).

In live sport settings, phygital tools are enhancing the fan experience. Venues like Levi's Stadium in California have implemented mobile apps that allow fans to order food, navigate the stadium, and access real-time player statistics, improving convenience and engagement on game day (Hurley, 2021). For sport-tech startups, these phygital strategies offer opportunities to expand reach, personalize services, and collect valuable user data to refine offerings, and by combining the physical experience of sport with the flexibility and connectivity of digital platforms, these startups are creating more immersive, scalable, and resilient solutions (Bonfanti et al., 2023).

Ethical Design, Public Health, and the Future of Sport Culture

The rise of quantified self-practices, enabled by startups offering wearables, tracking apps, and data platforms, has also normalized a new paradigm of fitness; one that emphasizes continuous monitoring, self-optimization, personalized feedback, and surveillance technologies. This degree of corporate intrusion into one's personal data has prompted researchers and social critics to be mindful of important privacy implications (Zuboff, 2019). Jones et al. (2022), for example, caution individuals about the challenges of normalizing surveillance technologies and handing oversensitive individual biometric data to private corporations. In this way, while these tools can serve to democratize health and fitness insights gleaned from personal technology, they also present risks such as over-reliance on algorithmic outputs or anxiety from constant performance tracking (Millington et al., 2025). Successful startups must navigate this tension by emphasizing education, human-centered design, and features that support long-term well-being over short-term performance gains, where partnerships with health professionals and sport scientists can lend credibility and guide ethical design.

In the realm of public health, sport-tech startups hold transformative potential by making fitness more engaging, personalized, and socially connected; these technologies can reduce barriers to regular physical activity. This is especially relevant among populations traditionally underrepresented in sport, for example, gamified apps have shown efficacy in increasing step counts, reducing sedentary behavior, and improving motivation among inactive individuals (Johnson et al., 2016). During pandemic-related lockdowns, startups played a vital role in helping users maintain activity levels despite gym closures (Seshadri et al., 2019), where these patterns suggest that digital sport platforms can serve as tools for population-level health promotion if access, usability, and affordability are addressed.

Finally, sport-tech startups are influencing the culture and values of sport itself. As digital and virtual formats become normalized, questions of sportspersonship, fairness, and community take on new meaning. In virtual cycling competitions, the boundaries of fair play are often ambiguous, raising the question of what precisely constitutes cheating in these digitally mediated environments (Richardson et al., 2023)? For sport-tech startups, success hinges not only on engineering functional platforms but also on cultivating a healthy digital culture. This entails designing norms, incentives, and user interfaces that prioritize inclusivity, transparency, and mutual respect.

Yet, there is also opportunity; by facilitating global, inclusive, and cross-cultural participation, digital sport-tech platforms can foster new forms of community. A beginner cyclist in Ghana can race alongside a professional in Germany; a retired marathoner in Boston can coach young runners in Manila. Startups that design for such connections—via social features, real-time communication, or shared challenges—are not just building products; they are creating global sport communities that redefine what it means to be an athlete in the 21st century.

Practice in Action

Zwift is a leading example of how sport technology startups can leverage gamification and virtual environments to enhance user engagement and motivation in fitness (Reed et al., 2023). Founded in 2014, Zwift is a virtual fitness platform that enables real-time, synchronous participation in cycling and running activities within immersive digital environments. By integrating with connected devices such as smart trainers and heart rate monitors, Zwift allows users to track performance metrics like speed, power, and cadence, while also controlling resistance to simulate real-world terrain. This combination of data-driven training and interactive, gamified worlds fosters a social and competitive atmosphere that enhances the indoor exercise experience (Smythe, 2025).

By 2020, the private company was rumored to be valued at around $1 billion following $450 million in a new investment round led by US private equity firm KKR & Co Inc. (Bhalla, 2020). Since that time, growth has steadily increased, and by January of 2025, Zwift had become one of the leading fitness and sport apps by revenue, ranking in the top eight worldwide (AppMagic, 2025).

Zwift's core innovation lies in its ability to create a virtual social community where users can ride or run alongside friends, compete in races, or participate in group workouts from anywhere. This social component, coupled with personalized training plans and structured challenges, promotes sustained engagement where users earn points, badges, and other rewards as they progress, reinforcing motivation through gamified elements (Ratten, 2021b). In a 2019 interview, when asked if Zwift was best understood as a fitness company or a gaming company, co-founder Eric Min explained that

> We're [Zwift] the fitness company born out of gaming. While we're a fitness brand, we're also a game and social network, two things that are converging rapidly right now. What we're trying to do, though, is build this social network around real-time experiences, physical experiences, and I think that's far more interesting.
>
> *(Peckham, 2019)*

Zwift exemplifies seamless integration of hardware and software by utilizing smart trainers and Bluetooth-enabled sensors to deliver real-time performance data. Metrics such as speed, cadence, power output, and heart rate are continuously tracked and displayed within the virtual environment, allowing users to monitor their progress moment-to-moment. This real-time access to data not only enhances user engagement but also supports the personalization of training through adaptive algorithms that adjust workout intensity based on current performance data. Such responsiveness transforms static workout routines into dynamic, data-informed experiences that promote sustained motivation and goal attainment. Furthermore, the app's ability to aggregate and analyze user data over time enables more nuanced insights into training trends, recovery needs, and performance plateaus—features increasingly vital in both amateur and competitive training contexts (Ometov et al., 2021). This integration of real-time feedback and data-driven adaptation underscores Zwift's appeal within the broader landscape of sport-tech innovation.

As Zwift continues to grow, it is actively exploring the integration of VR and AR technologies to deepen user immersion and engagement. These advancements could transform remote fitness experiences into highly interactive and socially connected virtual environments, further blurring the line between gaming and exercise (de la Hera, 2019). However, the implementation of such technologies also raises concerns. First, the demand for compatible hardware—such as advanced sensors, VR headsets, and powerful processors—may limit accessibility for users with fewer resources, potentially reinforcing digital divides in fitness technology (Millington & Wilson, 2020). Moreover, as Zwift collects and processes vast amounts of user data, including biometric and geolocation information, issues around data privacy and protection become increasingly pressing. Ensuring compliance with international data governance standards and maintaining user trust will be critical for sustained scalability and global adoption (Zuboff, 2019).

Summary

Sport-tech startups represent a transformative force in the modern sport and fitness landscape, reshaping how individuals train, compete, and engage with physical activity through the integration of cutting-edge technology. This chapter has traced the historical lineage of sport tech innovation and startups, from early 20th-century measurement tools to the current convergence of AI, biometric wearables, and immersive digital platforms. These ventures leverage the agility of startup culture to outpace traditional firms, delivering user-centric, data-driven solutions that personalize training, democratize access to performance analytics, and enhance both elite and recreational participation. Companies like Zwift, WHOOP, and Peloton illustrate how sport-tech startups are reimagining physical activity as a digitally enriched, socially connected, and highly adaptive experience.

At the core of this transformation lies a series of disruptive business models that emphasize recurring revenue, behavioral engagement, and ecosystem development. Startups no longer rely solely on equipment sales but offer ongoing value through subscriptions, gamified platforms, and real-time feedback loops. These models deepen user retention and foster long-term brand loyalty by creating communities around shared data, goals, and digital experiences. As seen in the rise of adaptive gamification and personalized feedback systems, startups are increasingly positioning themselves not just as fitness providers, but as long-term behavioral partners that support physical, mental, and social well-being. This shift has also aligned many ventures with broader public health goals, opening new avenues for cross-sector partnerships and impact.

The growth of sport technology is accompanied by critical ethical, legal, and societal implications (Studnicka, 2020). The increasing integration of biometric monitoring and algorithmic decision-making in athletic contexts raises urgent concerns related to data privacy, surveillance, and digital equity

(Berg, 2024). Regulatory oversight remains inconsistent and often inadequate, leaving sport-tech startups to navigate a fragmented and evolving legal landscape. Emerging issues such as technological doping, the commercialization of athlete data, and algorithmic bias further threaten the integrity and inclusivity of digital sport experiences (Petersen et al., 2025). As emphasized in this chapter, sport-tech startups must move beyond technological novelty by embedding ethical design and operational strategies at their core. Only by prioritizing trust, transparency, and fairness can these ventures ensure long-term sustainability in an increasingly data-driven sport ecosystem.

Ultimately, sport-tech startups are not merely responding to changes in sport and society—they are active agents shaping the future of physical activity, health, and digital culture. Their innovations are blurring the lines between virtual and physical spaces, altering the definitions of athleticism and participation, and forging new global communities through phygital experiences. As the sector matures, stakeholders across academia, industry, and policy will need to engage collaboratively with these ventures to foster responsible innovation. By understanding the startup ecosystem, the forces driving its growth, and the values guiding its evolution, we can better anticipate and shape the role of sport-tech in a rapidly digitizing world.

Key Resources

- **Zwift**: www.zwift.com
- **Freeletics**: https://www.freeletics.com
- **Strava**: https://www.strava.com
- **Peloton**: https://www.onepeloton.com/
- **Dartfish**: https://www.dartfish.com/

References

Allouch, M. (2024). Democratizing performance: Impact of the data revolution on recreational running (Doctoral dissertation). Retrieved from https://dspace.mit.edu/handle/1721.1/155998

AppMagic. (March 5, 2025). Leading fitness and sport apps worldwide in January 2025, by revenue (in million U.S. dollars) [Graph]. *Statista*. Retrieved May 28, 2025, from https://www.statista.com/statistics/1239716/top-fitness-and-sport-apps-by-revenue/

Arnow, G. (2016). Apple watching you: Why wearable technology should be federally regulated. *Loyola Los Angeles Law Review*, *49*, 607.

Aughey, R. J. (2011). Applications of GPS technologies to field sports. *International Journal of Sports Physiology and Performance*, *6*(3), 295–310.

Baena-Arroyo, M. J., García-Fernández, J., Gálvez-Ruiz, P., & Grimaldi-Puyana, M. (2020). Analyzing consumer loyalty through service experience and service convenience: differences between instructor fitness classes and virtual fitness classes. *Sustainability*, *12*(3), 828.

Batat, W. (2022). What does phygital really mean? A conceptual introduction to the phygital customer experience (PH-CX) framework. *Journal of Strategic Marketing*, *32*(8), 1220–1243. https://doi.org/10.1080/0965254X.2022.2059775

Bates, S., Greene, D., & O'Quinn, L. (2021). Virtual sport-based positive youth development during the COVID-19 pandemic. *Child and Adolescent Social Work Journal*, *38*(4), 437–448.

Behzadnia, B., Alizadeh, E., Haerens, L., & Aghdasi, M. T. (2022). Changes in students' goal pursuits and motivational regulations toward healthy behaviors during the pandemic: A self-determination theory perspective. *Psychology of Sport and Exercise*, *59*, 102131.

Berg, J. V. (2024). The game-changer: Legal issues surrounding wearable technology in sports. *SMU Science & Technology Law Review*, *27*, 151.

Berg, S. (2022). Data privacy and security in the digital fitness industry. *Journal of Sport Tech*, *15*(3), 45–58.

Berg, S., & Fischer, T. (2023). The monetization of sports technology and the role of data analytics. *Journal of Sport Management & Innovation*, *11*(1), 92–107.

Bhalla, A. (2020, September 16). Online fitness training platform Zwift raises $450 million in funding round led by KKR. *Reuters*. Retrieved from https://www.reuters.com/article/business/online-fitness-training-platform-zwift-raises-450-million-in-funding-round-led-idUSKBN2671PS/

Bonfanti, A., Vigolo, V., Vannucci, V., & Brunetti, F. (2023). Creating memorable shopping experiences to meet phygital customers' needs: Evidence from sporting goods stores. *International Journal of Retail & Distribution Management*, *51*(13), 81–100.

Bouhaj, S., Jahidi, R., & Lebzar, B. (2022). A conceptual and semantic framework of the startup: A systematic review of the literature through a qualitative study. *Revue Internationale du Chercheur*, *3*(2), 39–51.

Business Insider. (2025, February). 10 sports startups that raised the most money from VCs last year. Retrieved from https://www.businessinsider.com/top-sports-startups-raised-millions-vc-last-year-unrivaled-lovb-2025-2

Butler, D. (2000). *Business Planning: A Guide to Business Start-Up*. Routledge.

Campo-Beamud, C., & Rodrigo-Carranza, V. (2024). Sport breakthroughs driven by scientific knowledge: A history of scientific evolution. *Journal of Applied Physiology*, *137*(3), 665–665.

Chopra, A., Singhal, A., & Understanding the Wearable Technology. (2021, April). *Proceedings of the International Conference on Innovative Computing & Communication (ICICC) 2021*. Retrieved from https://papers.ssrn.com/sol3/papers.cfm?abstract_id=3833316

Choudhary, A. (2024). Internet of things: A comprehensive overview, architectures, applications, simulation tools, challenges and future directions. *Discover Internet of Things*, *4*(1), 31.

Clark, R. (2021). The future of sport tech: Professionalization, data privacy, and accessibility. *International Journal of Sport Technology*, *19*(4), 210–225.

Cohen, S., & Hochberg, Y. V. (2014). Accelerating startups: The seed accelerator phenomenon. *SSRN Electronic Journal*. https://doi.org/10.2139/ssrn.2418000

Cossich, V. R. A., Carlgren, D., Holash, R. J., & Katz, L. (2023). Technological breakthroughs in sport: Current practice and future potential of artificial intelligence, virtual reality, augmented reality, and modern data visualization in performance analysis. *Applied Sciences*, *13*(23), 12965. https://doi.org/10.3390/app132312965

Crawford, K., Lingel, J., & Karppi, T. (2015). Our metrics, ourselves: A hundred years of self-tracking from the weight scale to the wrist wearable device. *European Journal of Cultural Studies*, *18*(4–5), 479–496.

de la Hera, T. (2019). *Digital gaming and the advertising landscape*. Amsterdam University Press.

Deloitte. (2021). Digital transformation in sports: From athlete performance to fan engagement. *Deloitte Insights*. Retrieved from https://www.deloitte.com/global/en/Industries/tmt/perspectives/gx-digital-transformation-and-future-changes-in-sports-industry.html

Edney, S., Olds, T., Ryan, J. C., Vandelanotte, C., Plotnikoff, R., Curtis, R., … Maher, C. (2019). A social networking and gamified app to increase physical activity: Cluster RCT. *American Journal of Preventive Medicine*, *56*(4), 464–474.

Esmer, Y., & Dertli, Ş. (2024). Bibliometric analysis of sport entrepreneurship literature: The case of Web of Science. *Research in Sport Education and Sciences*, *26*(4), 159–176.

Espinosa, H., Mears, A., Stamm, A., Ohgi, Y., & Coniglio, C. (2025). Wearable sensor technology in sports monitoring. *Sports Engineering*, *28*, 4. https://doi.org/10.1007/s12283-025-00485-9

Farhan, K. A., Asadullah, A. B. M., Kommineni, H. P., Gade, P. K., & Venkata, S. S. M. G. N. (2023). Machine learning-driven gamification: Boosting user engagement in business. *Global Disclosure of Economics and Business*, *12*(1), 41–52.

Ferreira, J. J., Fernandes, C., Ratten, V., & Miragaia, D. (2020). Sports innovation: A bibliometric study. In V. Ratten (Ed.), *Sport Entrepreneurship and Public Policy: Building a New Approach to Policy-Making for Sport* (pp. 153–170). Springer International Publishing.

Flynn, J. (2024). *Sports and Technology Have the Power to Change the World: Driving Positive Change Through the Use of Data and AI*. John Wiley & Sons.

Frevel, N., Schmidt, S. L., Beiderbeck, D., Penkert, B., & Subirana, B. (2023). Taxonomy of sportstech. In S. Schmidt (Ed.), *21st Century Sports: How Technologies Will Change Sports in the Digital Age* (pp. 17–39). Springer International Publishing.

García-Peñalvo, F. J. (2021). Avoiding the dark side of digital transformation in teaching. An institutional reference framework for eLearning in higher education. *Sustainability*, *13*(4), 2023.

Gard, M., Dionigi, R. A., & Horton, S. (2020). Surveillance, digital technology and youth sport: Policy and practice. *Sport, Education and Society*, *25*(2), 132–146.

Gazzola, P., Pavione, E., & Ferrazzano, F. (2024). *Evolution of the Global Fitness Industry: Strategy, Sustainability and Innovation*. Taylor & Francis.

Ghosh, I., Ramamurthy, S. R., Chakma, A., & Roy, N. (2023). Sports analytics review: AI applications, emerging technologies and algorithmic perspective. *Wiley Interdisciplinary Reviews: Data Mining and Knowledge Discovery*, *13*(5), e1496. https://doi.org/10.1002/widm.1496

Glebova, E., & Desbordes, M. (2021). Technology innovations in sport: Typology, nature, courses and impact. In P. Pedersen (Ed.), *Innovation and Entrepreneurship in Sport Management*, (pp. 57–72). Edward Elgar Publishing.

Glebova, E., Lee, H., & Kim, M. (2024). The impact of pandemic-induced digitization on fitness and sports technology adoption. *Sport Management Review*, *27*(1), 22–35.

Glebova, E., Madsen, D. Ø., Mihaľová, P., Géczi, G., Mittelman, A., & Jorgič, B. (2024). Artificial intelligence development and dissemination impact on the sport industry labor market. *Frontiers in Sport and Active Living*, *6*, 1363892.

Glebova, E., Su, Y., Desbordes, M., & Schut, P.-O. (2025). Editorial: Emerging digital technologies as a game changer in the sport industry. *Frontiers Sports Active and Living*, *7*, 1605138.

Global Sports Innovation Center (GSIC). (2019). Digital transformation of sports entities by 2025. Retrieved from https://sport-gsic.com/wp-content/uploads/2019/07/GSIC-Report-Digital-Transformation-of-Sports-Entitites-by-2025-VL.pdf

Gompers, P., & Lerner, J. (2001). The venture capital revolution. *Journal of Economic Perspectives*, *15*(2), 145–168.
González-Serrano, M. H., Añó Sanz, V., & González-García, R. J. (2020). Sustainable sport entrepreneurship and innovation: A bibliometric analysis of this emerging field of research. *Sustainability*, *12*(12), 5209. https://doi.org/10.3390/su12125209
González-Serrano, M. H., Valantine, I., & Hervás, J. C. (2020). Sport entrepreneurship: A bibliometric and visualization analysis. *International Entrepreneurship and Management Journal*, *16*(3), 789–812. https://doi.org/10.1007/s11365-019-00606-5
Grimaldi, R., & Grandi, A. (2005). Business incubators and new venture creation: An assessment of incubating models. *Technovation*, *25*(2), 111–121. https://doi.org/10.1016/S0166-4972(03)00076-2
Hajarian, M., Bastanfard, A., Mohammadzadeh, J., & Khalilian, M. (2019). A personalized gamification method for increasing user engagement in social networks. *Social Network Analysis and Mining*, *9*, 1–14.
Hajli, N. (2025). Exploring fan engagement and value co-creation in virtual sport communities: A comparative study of virtual sport communities over time. *Technological Forecasting and Social Change*, *212*, 123934.
Hammerschmidt, J., González-Serrano, M. H., Puumalainen, K., & Calabuig, F. (2024). Sport entrepreneurship: The role of innovation and creativity in sport management. *Review of Managerial Science*, *18*(11), 3173–3202.
Huang, M., & Yongquan, T. (2025). Tech-driven excellence: A quantitative analysis of cutting-edge technology impact on professional sports training. *Journal of Computer Assisted Learning*, *41*(1), e13082.
Huang, X. (2022, October). Data visualization design strategies for promoting exercise motivation in self-tracking applications. In *Proceedings of the 40th ACM International Conference on Design of Communication, Lisbon, Portugal* (pp. 78–89).
Huertas González-Serrano, M., Jones, P., & Llanos-Contrera, O. (2019). An overview of sport entrepreneurship field: A bibliometric analysis of the articles published in the Web of Science. *Sport in Society*, *23*(2), 296–314. https://doi.org/10.1080/17430437.2019.1607307
Hurley, O. A. (2021). Sport cyberpsychology in action during the COVID-19 pandemic (opportunities, challenges, and future possibilities): A narrative review. *Frontiers in Psychology*, *12*, 621283.
Hutton, E., (2021). Peloton Interactive, Inc.: Valuation during a global pandemic. In *Sage Business Cases*. SAGE Publications, Ltd. https://doi.org/10.4135/9781529764185
Jessop, A., & Baker, T. A. III. (2019). Big data bust: Evaluating the risks of tracking NCAA athletes' biometric data. *Texas Review of Entertainment & Sports Law*, *20*(1), 81–112.
Johnson, D., Deterding, S., Kuhn, K. A., Staneva, A., Stoyanov, S., & Hides, L. (2016). Gamification for health and wellbeing: A systematic review of the literature. *Internet Interventions*, *6*, 89–106.
Jones, L., Konoval, T., & Toner, J. (2022). Sport and surveillance technologies. In *Sport, Social Media, and Digital Technology: Sociological Approaches*, ed. Jimmy Sanderson (pp. 165–183). Emerald Publishing Limited.
Jones, P., Ratten, V., & Hayduk, T. (2020). Sport, fitness, and lifestyle entrepreneurship. *International Entrepreneurship and Management Journal*, *16*, 783–793.
Lad, L. J., & Caldwell, C. B. (2009). Collaborative standards, voluntary codes and industry self-regulation: The role of third-party organizations. *Journal of Corporate Citizenship*, 35, 67–78. https://www.jstor.org/stable/jcorpciti.35.67
Laukkanen, R. M. T., & Virtanen, P. K. (1998). Heart rate monitors: State of the art. *Journal of Sports Sciences*, *16*(1), 3–7. https://doi.org/10.1080/026404198366920
Li, A., & Huang, W. (2024). A comprehensive survey of artificial intelligence and cloud computing applications in the sport industry. *Wireless Networks*, *30*(8), 6973–6984.

Lokey, H. (2024). *Sports Market Update Fall 2024*. Retrieved from https://www2.hl.com/pdf/2024/sports-market-update-fall-2024.pdf

Maguire, J. (2015). Social media, sport, and the transformation of athlete-fan interactions: A sociological analysis. *Sport in Society*, *18*(5), 553–564.

Malik, P., Shahdadpuri, P., Malviya, R., Patil, R., Badwaniya, V., & Sthapak, R. (2024). Biometric data analysis in athlete monitoring: Advancements, applications, and implications. *International Journal of Innovative Research in Technology*, *10*(12), 2882–2889.

Mann, S., Smith, T., & Krishnan, R. (2020). Navigating the GDPR in health-related apps and wearables: Legal and design strategies. *Health and Technology Law Review*, *3*(1), 1–17.

Martín-Peña, M. L., Lorenzo, P. C., & Meyer, N. (2024). Digital platforms and business ecosystems: A multidisciplinary approach for new and sustainable business models. *Review of Managerial Science*, *18*(9), 2465–2482.

MasterBorn. (2023, October 10). The 10 top innovative sports tech startups you should know. Retrieved from https://www.masterborn.com/blog/the-10-top-innovative-sports-tech-startups

Mazzucato, M. (2013). *The Entrepreneurial State: Debunking Public vs. Private Sector Myths*. Anthem Press.

McChesney, R. W. (2004). *The Problem of the Media: U.S. Communication Politics in the Twenty-First Century*. NYU Press.

Mele, C., Spena, T. R., Marzullo, M., & Di Bernardo, I. (2023). The phygital transformation: A systematic review and a research agenda. *Italian Journal of Marketing*, *2023*(3), 323–349.

Methlagl, M., Mairhofer, S., & Michlmayr, F. (2025). Exploring technology acceptance patterns of users of the mixed-reality sport technology Zwift: Antecedents and consequences of technology acceptance. *Universal Access in the Information Society*, *24*, 721–732. https://doi.org/10.1007/s10209-024-01094-2

Miller, K. M., Lukic, K., & Skiera, B. (2024). The impact of the general data protection regulation (GDPR) on online tracking. *arXiv preprint arXiv:2411.06862*.

Millington, B., Naraine, M. L., Wanless, L., Safai, P., & Manley, A. (2025). Sport and the promise of artificial intelligence: Human and machine futures. *Sociology of Sport Journal*, *1*, 1–10.

Millington, B., & Wilson, B. (2020). *The Greening of Golf: Sport, Global Warming and the Politics of Sustainability*. Manchester University Press.

Motti, V. G. (2020). Introduction to wearable computers. In *Wearable Interaction. Human–Computer Interaction Series*. Springer. https://doi.org/10.1007/978-3-030-27111-4_1

Najjar, M. C. (2023). Legal and ethical issues arising from the application of data analytics and artificial intelligence to traditional sports. *Albany Law Journal Science & Technology*, *33*, 51.

Napoli, P. M. (2022). Social media and the shift to subscription models in digital fitness: Consumer behavior and platform dynamics. *Journal of Media Economics*, *35*(1), 23–39.

Nithya, N., & Nallavan, G. (2023). Monitoring biometric data of a player using a wearable device in real time for sports applications. *Wireless Personal Communications*, 133, 981–993. https://doi.org/10.1007/s11277-023-10800-x

Oliveira, F., Costa, D. G., Assis, F., & Silva, I. (2024). Internet of intelligent things: A convergence of embedded systems, edge computing and machine learning. *Internet of Things*, 26, 101153. https://doi.org/10.1016/j.iot.2024.101153

Ometov, A., Shubina, V., Klus, L., Skibińska, J., Saafi, S., Pascacio, P., … Lohan, E. S. (2021). A survey on wearable technology: History, state-of-the-art and current challenges. *Computer Networks*, *193*, 108074.

Peckham, E. (2019, April 25). Zwift CEO Eric Min on how the startup is turning indoor cycling into an esport. *TechCrunch*. https://techcrunch.com/2019/04/25/interview-zwift-ceo-eric-min/

Petersen, T. S., Holmen, S. J., & Ryberg, J. (2025). AI, doping and ethics: On why increasing the effectiveness of detecting doping fraud in sport may be morally wrong. *Journal of Medical Ethics*, *51*(2), 102–106.

Pickering, C., & Kiely, J. (2019). The development of a personalised training framework: Implementation of emerging technologies for performance. *Journal of Functional Morphology and Kinesiology*, *4*(2), 25.

Pizzo, A. D., Baker, B. J., Jones, G. J., & Funk, D. C. (2020). Sport experience design: Wearable fitness technology in the health and fitness industry. *Journal of Sport Management*, 35(2), 130–143. https://journals.humankinetics.com/view/journals/jsm/35/2/article-p130.xml

Qi, Y., Sajadi, S. M., Baghaei, S., Rezaei, R., & Li, W. (2024). Digital technologies in sports: Opportunities, challenges, and strategies for safeguarding athlete wellbeing and competitive integrity in the digital era. *Technology in Society*, *77*. https://doi.org/10.1016/j.techsoc.2024.102496

Ratten, V. (2019). *Sport Technology and Innovation*. Springer.

Ratten, V. (2020a). Digital transformation in sport and social media. In V. Ratten (Ed.), *Sport Startups: New Advances in Entrepreneurship* (pp. 89–104). Emerald Publishing Limited.

Ratten, V. (2020b). Entrepreneurial ecosystems: The role of accelerators. In V. Ratten (Ed.), *Entrepreneurship as Empowerment: Knowledge Spillovers and Entrepreneurial Ecosystems* (pp. 11–22). Emerald Publishing Limited.

Ratten, V. (2020c). Sport startups: What does the future hold? In V. Ratten (Ed.), *Sport Startups: New Advances in Entrepreneurship* (pp. 89–104). Emerald Publishing Limited.

Ratten, V. (2021a). Introduction: Innovation and entrepreneurship in sport management. In V. Ratten (Ed.),*Innovation and Entrepreneurship in Sport Management* (pp. 1–8). Edward Elgar Publishing. https://doi.org/10.4337/9781783473960.00008

Ratten, V. (2021b). Sport entrepreneurial ecosystems and knowledge spillovers. *Knowledge Management Research & Practice*, *19*(1), 43–52.

Ratten, V. (2023). Entrepreneurship: Definitions, opportunities, challenges, and future directions. *Global Business and Organizational Excellence*, *42*(5), 79–90.

Ratten, V. (2024). *Sport Entrepreneurial Ecosystems*. Springer.

Ratten, V., & Jones, P. (2020). New challenges in sport entrepreneurship for value creation. *International Entrepreneurship and Management Journal*, *16*(3), 961–980.

Ratten, V., & Thompson, A. J. (2020). Digital sport entrepreneurial ecosystems. *Thunderbird International Business Review*, *62*(5), 565–578.

Reed, J., Dunn, C., Beames, S., & Stonehouse, P. (2023). E 'Ride on!': The Zwift platform as a space for virtual leisure. *Leisure Studies*, *42*(2), 188–202.

Reischauer, G., Hess, T., Sellhorn, T., & Theissen, E. (2024). Transparency in an age of digitalization and responsibility. *Schmalenbach Journal of Business Research*, *76*, 483–494. https://doi.org/10.1007/s41471-024-00203-4

Richardson, A., Berger, N., & Smith, P. (2023). Perceptions of cheating and doping in E-cycling. *Sports*, *11*(10), 201.

Saltarella, M., Desolda, G., & Lanzilotti, R. (2021). Privacy design strategies and the GDPR: A systematic literature review. In G. Meiselwitz (Ed.), *HCI for Cybersecurity, Privacy and Trust: HCII 2021. Lecture Notes in Computer Science* (Vol. 12786, pp. 241–257). Springer. https://doi.org/10.1007/978-3-030-77392-2_16

Sampedro, A. C. (2023). *The Ethics of AI in Sport. Taking Athletes' Rights and Wellbeing Seriously*. Final report, The IOC Olympic Studies Centre.

Sardi, L., Idri, A., & Fernández-Alemán, J. L. (2017). A systematic review of gamification in e-health. *Journal of Biomedical Informatics*, *71*, 31–48. https://doi.org/10.1016/j.jbi.2017.05.011

Seshadri, D. R., Drummond, C., Craker, J., Rowbottom, J. R., & Voos, J. E. (2017). Wearable devices for sports: new integrated technologies allow coaches, physicians, and trainers to better understand the physical demands of athletes in real time. *IEEE Pulse*, *8*(1), 38–43. https://ieeexplore.ieee.org/document/7831543

Seshadri, D. R., Li, R. T., Voos, J. E., Rowbottom, J. R., Alfes, C. M., Zorman, C. A., & Drummond, C. K. (2019). Wearable sensors for monitoring the internal and external workload of the athlete. *NPJ Digital Medicine*, *2*(1), 71.

Sethi, S., & Talvelkar, S. (2024). A comprehensive review of emerging trends in sports management. *Multidisciplinary Reviews*, *7*(4), 2024068.

Sifaoui, A., & Eastin, M. S. (2024). "Whispers from the wrist": Wearable health monitoring devices and privacy regulations in the US: The loopholes, the challenges, and the opportunities. *Cryptography*, *8*(2), 26.

Sinar, M., & Acar, M. (2024a). The role of biosensor data in athlete performance optimization. *Journal of Sports Technology and Science*, *11*(2), 45–59.

Sinar, M., & Acar, M. (2024b). Wearable sports technologies. *International Journal of Sports Engineering and Biotechnology*, *2*(1), 17–26.

Singh, R., & Khan, A. (2022). Marketing strategies of sports tech startups: Digital channels and influencer partnerships. *Journal of Digital Marketing*, *17*(2), 78–94.

Smythe, S. (2025, January 8). Eric Min interview: Why cycling in London drove me to create Zwift – and what Watopia has taught me in the 10 years since. *Rouleur*. https://www.rouleur.cc/blogs/desire-journal/eric-min-interview-why-living-in-london-drove-me-to-create-zwift-and-how-it-went-global

Statista. (2024, September 19). Number of fitness tracker users in the United States from 2017 to 2029, by segment (in millions) [Graph]. *Statista*. Retrieved May 23, 2025, from https://www.statista.com/forecasts/1425172/fitness-tracker-users-by-segment-us

Studnicka, A. (2020). The emergence of wearable technology and the legal implications for athletes, teams, leagues and other sports organizations across amateur and professional athletics. *DePaul Journal of Sports Law*, *16*, i.

Supriyanto, A., & Liu, Y. (2022). 'Ride on!': The Zwift platform as a space for virtual leisure. *Leisure Studies*, *41*(4), 499–512. https://doi.org/10.1080/02614367.2022.2088836

SVG Europe. (2023, March). Artificial intelligence: Ethical considerations in emerging sports entertainment technology. Retrieved from https://www.svgeurope.org/blog/headlines/artificial-intelligence-ethical-considerations-in-emerging-sports-entertainment-technology/

Toohey, K., & Veal, A. J. (2007). *The Olympic Games: A Social Science Perspective*. CABI .

Troise, C., Matricano, D., Candelo, E., & Sorrentino, M. (2022). Entrepreneurship and fintech development: comparing reward and equity crowdfunding. *Measuring Business Excellence*, *I*(1), 52–63.

UCI adds eSports World Championships in 2020. (2019, September 26). *Cyclingnews*. Retrieved from https://www.cyclingnews.com/news/uci-adds-esports-world-championships-in-2020/

Wang, X. (2021). The pandemic's impact on the sport tech industry: An analysis of the rise of home fitness startups. *International Journal of Sport Innovation*, *6*(2), 112–128.

Weiblen, T., & Chesbrough, H. W. (2015). Engaging with startups to enhance corporate innovation. *California Management Review*, *57*(2), 66–90. https://doi.org/10.1525/cmr.2015.57.2.66

Westmattelmann, D., Grotenhermen, J. G., Sprenger, M., & Schewe, G. (2021). Apart we ride together: The motivations behind users of mixed-reality sports platforms. *Journal of Business Research*, *134*, 663–672.

Wetzel, W. E. (1983). Angels and informal risk capital. *Sloan Management Review*, *24*(4), 23–34.

WHOOP. (2019, January 2). *Membership*. Retrieved from https://web.archive.org/web/20190102193724/

Wilke, J., Mohr, L., Tenforde, A. S., Edouard, P., Fossati, C., González-Gross, M., … Hollander, K. (2021). A pandemic within the pandemic? Physical activity levels substantially decreased in countries affected by COVID-19. *International Journal of Environmental Research and Public Health*, *18*(5), 2235.

Winchester, C. C., Pleggenkuhle-Miles, E., & Bass, A. E. (2021). Peloton's ride to growth. *The CASE Journal*, *17*(6), 754–783.

Xiao, L., Cao, Y., Gai, Y., Liu, J., Zhong, P., & Moghimi, M. M. (2023). Review on the application of cloud computing in the sports industry. *Journal of Cloud Computing*, *12*(1), 152. https://doi.org/10.1186/s13677-023-00531-6

Xiao, L., Zhang, Y., & Yilmaz, Y. (2023). Artificial intelligence and IoT in sports: Trends and future directions. *IEEE Internet of Things Journal*, *10*(1), 112–126.

Yan, T., & Zhang, C. (2022). Active fairness auditing. In K. Chaudhuri, S. Jegelka, L. Song, C. Szepesvári, G. Niu, & S. Sabato (Eds.), *Proceedings of the 39th International Conference on Machine Learning* (Vol. 162, pp. 24929–24962). Proceedings of Machine Learning Research (PMLR).

Zuboff, S. (2019). *The Age of Surveillance Capitalism: The Fight for a Human Future at the New Frontier of Power*. Public Affairs.

Zuckerman, O., & Gal-Oz, A. (2014). Deconstructing gamification: Evaluating the effectiveness of continuous measurement, virtual rewards, and social comparison for promoting physical activity. *Personal and Ubiquitous Computing*, *18*(7), 1705–1719.

PART III

Social and Community Engagement in Digital Spaces

10

DIGITAL COMMUNITIES IN SPORT

Gashaw Abeza and Jacob Bustad

Introduction

Historically, sports communities were primarily local, characterized by in-person interactions among fans gathered in stadiums, bars, or community centers to collectively enjoy sporting events (Gantz & Lewis, 2014; Stewart et al., 2003). However, technological advancements have reshaped these communities considerably, facilitating the shift from local, face-to-face interactions to digital, globally connected networks (Yadav et al., 2023). Modern digital platforms, including social media, online forums, mobile apps, and immersive technologies, have allowed sport enthusiasts to form interactive online communities beyond geographical constraints (Hull & Abeza, 2021). This evolution has significantly expanded opportunities for cocreation of fan experiences and content, increasing global reach and brand awareness, digital sales, fan engagement and retention, as well as collective participation in sport culture (Abeza et al., 2017; Wagner, 2021).

A digital community involves interactions mediated by diverse digital technologies and platforms (Yadav & Pavlou, 2020), including mobile applications, social media, augmented reality (AR), and virtual reality (VR). Communities on and around these platforms leverage advanced digital tools to enhance user engagement, facilitate ongoing interactions, and promote the formation of collective identities (Ströbel et al., 2021). By removing geographical barriers, digital communities enhance global connectivity, allow real-time interactions, and facilitate collaborative content creation, fostering stronger emotional ties and a deeper sense of belonging among participants (Glebova et al., 2025). Consequently, these communities have transformed traditional sports engagement, promoting greater participation, shared interest, loyalty, and interactive fan experiences (Abeza & King-White, 2024).

DOI: 10.4324/9781003588405-13

This chapter specifically explores various types of digital communities in contemporary sport contexts, such as esports, fantasy sports, virtual fitness groups, running apps, AR and VR communities, social media influencer groups, football club fan forums, and Facebook groups. Each community's unique characteristics, operational mechanisms, and illustrative examples will be discussed. Furthermore, the chapter includes practical cases and highlights current trends and issues, underscoring the practical and academic relevance of understanding digital communities within sport management research and practice.

Background

Historical Evolution of Digital Communities

Historically, sport fandom was primarily localized, characterized by face-to-face interactions and shared cultural practices (Stewart et al., 2003). Fans typically gathered in stadiums, local bars, community centers, and public viewing spaces to experience live sporting events collectively, creating environments conducive to identity formation and community belonging (Gantz & Lewis, 2014). The reliance on traditional print media and word of mouth reinforced the localized nature of sport fandom, with individuals accessing news, commentary, and analysis mainly within their immediate geographical and social contexts (Rowe, 2011). However, the late twentieth century marked a shift driven by rapid advancements in digital technology, including the emergence and global use of personal computers, widespread internet connectivity, and the advent of portable devices (Hull & Abeza, 2021). These advancements facilitated the transition from geographically bound interactions to globally accessible digital forums, enabling sport enthusiasts to transcend traditional spatial limitations and form expansive, diverse communities centered around shared sporting interests (Abeza et al., 2017).

In the late 1980s and early 1990s, digital sport communication emerged through platforms such as Bulletin Board Systems (BBS) and Usenet (Boyd & Ellison, 2007), which provided initial online gathering spaces where fans could discuss game strategies, share experiences, and analyze sporting events asynchronously (Hutchins & Rowe, 2012). The threaded discussion format of these systems enabled continuous and reflective interactions, demonstrating the potential of digital environments to replicate and enhance communal aspects traditionally associated with sport fandom (Hutchins & Rowe, 2012; Wagner, 2021). Subsequently, the early 2000s saw the rise of personal and group blogs, significantly democratizing sport journalism (Hull & Abeza, 2021). Blogs allowed fans to express their views, critique mainstream sport narratives, and engage with personalized and subjective commentary, creating a more interactive and participatory model of content creation that influenced established media practices (Kian et al., 2011).

The most significant shift in digital sport communication took place from the mid-2000s onward with the advent of social media platforms such as Facebook, Twitter, Instagram, and TikTok. These platforms fundamentally altered the dynamics of sport fandom by enabling real-time, multidirectional communication among fans, athletes, and organizations (Hull & Abeza, 2021). Social media facilitates a participatory culture in which content generation and consumption blur traditional boundaries between producers and audiences, giving rise to phenomena such as live-tweeting and viral fan-driven campaigns that mobilize and engage large-scale communities instantaneously (Abeza et al., 2017). Today, advancements in digital technologies such as high-speed internet, portable smart devices, mobile computing, and cloud services have fostered environments characterized by real-time data dissemination and interactive content (Hull & Abeza, 2021). Such technological developments have enabled the formation of digital communities that transcend geographical limitations. Platforms and communities such as Facebook groups, virtual fitness communities, fantasy sports, esports forums, Reddit communities, and running app groups democratize content creation and consumption, allowing fans to act as co-creators of sport narratives (Kim & Kim, 2017). Consequently, digital communities facilitate decentralized community building, with individuals connected by common goals, interests, and needs, fostering collective identity, emotional attachment, and loyalty toward specific sport entities (Abeza et al., 2017; Kim & Kim, 2017).

In academic discourse, scholars have used various terms to describe community formation within mediated environments, including digital communities (Yadav & Pavlou, 2020), brand communities (Constantin et al., 2014; Wagner, 2023), online communities (Hammond, 2017; Wagner, 2021), and virtual communities (Porter, 2004). To establish conceptual clarity, this chapter explicitly addresses digital communities in the context of sport, with subsequent sections elaborating on conceptual variations, operational features, and illustrative real-world cases.

Topical Exploration and Key Concepts

Conceptual Variations of Community in Digital Contexts

As noted above, the emergence of digital technologies has led scholars to propose various concepts to better understand how communities emerge and function within mediated settings (e.g., Constantin et al., 2014; Hammond, 2017; Porter, 2004; Yadav & Pavlou, 2020). These conceptualizations include brand communities, online communities, digital communities, and virtual communities, each denoting distinct modes of interaction and affiliation. To establish clarity and set the focus of this chapter, it is important to differentiate among these conceptual variations of community in digital contexts.

As discussed below, each represents a distinct mode of interaction and affiliation, contributing uniquely to the construction of connection, loyalty, and engagement within the sport context.

A *digital community* involves interactions mediated by diverse digital technologies and platforms (Yadav & Pavlou, 2020), including mobile applications, social media, AR, and VR. Communities using these technologies and platforms often leverage sophisticated digital tools to foster deeper engagement. For example, many virtual fitness communities use AR or wearable devices to provide interactive user experiences (Rauschnabel, 2021). Related, yet distinct, a *virtual community* refers to a group whose interactions occur within computer-mediated or immersive digital spaces, frequently integrating aspects of both physical and digital experiences (Porter, 2004). For example, participants in FIFA eSports leagues engage within fully virtual environments, interacting entirely through digital interfaces and platforms (Hamari & Sjöblom, 2017). While virtual communities often share characteristics with other community forms, such as brand communities or online communities, they also highlight brand affiliation and offer digitally mediated interactions or immersive virtual environments (Muniz & O'Guinn, 2001; Yadav & Pavlou, 2020).

In contrast, an *online community* is broadly considered as a group of individuals who interact digitally around a shared interest, purpose, or activity, without necessarily being tied to a specific brand (Dover & Kelman, 2018; Wagner, 2021). These communities exist primarily through internet-based interactions that take place on various platforms, such as forums, social media groups, and discussion boards. Reddit serves as a prominent example, hosting numerous subreddits, each of which is dedicated to specific topics ranging from hobbies and interests to social causes that form the nucleus of discussions, shared content, and foundational connections (Ovadia, 2015).

On the other hand, a *brand community*, according to Muniz and O'Guinn (2001), refers to a specialized, non-geographically bound group of individuals connected by a shared passion for, and identification with, a specific brand. Members interact around brand-related experiences, products, or services, fostering loyalty and advocacy. For instance, the Nike Run Club exemplifies a brand community in which participants bond through their mutual attachment to Nike products and shared running experiences (Rodrigues et al., 2021). According to the authors, the Nike Run Club application employs gamification strategies to enhance user engagement and brand attitude, highlighting the role of social interaction in fostering a sense of community among users.

Beyond the four conceptual variants, additional community forms, such as fan communities and social media communities, are also important in helping establish conceptual clarity and focus of the chapter. In this regard, a *social media community* is a group of individuals who interact primarily through social media platforms, such as Facebook, Instagram, TikTok, and X (formerly Twitter) (Abeza et al., 2019). As the authors have noted, members of these communities are typically connected by shared content, topics, hashtags, or common interests in influential social media personalities, usually referred to as influencers. In contrast, a fan community is characterized by individuals who share a collective emotional investment in a sports team, athlete, or organization, engaging in both physical and digital environments. These communities are underpinned by shared rituals, symbols, and narratives that foster a sense of belonging and identity among members (Wann et al., 2001). For instance, supporters of Arsenal Football Club participate in various forms of engagement, including attending matches at the Emirates Stadium, interacting on social media platforms such as Twitter, and contributing to fan-led media outlets such as AFTV (Arsenal Fan TV), which provide spaces for expression and community building (Gibbons, 2019). The integration of digital platforms has expanded the reach and depth of fan communities, allowing for continuous interaction and the reinforcement of group identity beyond physical gatherings (Sanderson & Kassing, 2011).

Given the conceptual variations of community in digital contexts, it is important to recognize that the term digital communities in sport serves as an overarching framework that encompasses brand communities, online communities, virtual communities, fan communities, and social media communities within sport contexts. Although each form reflects distinct dynamics, they are fundamentally enabled, shaped, impacted, or mediated through digital technologies. Therefore, this chapter reflects the broad and evolving landscape of how sport fandom, participation, engagement, and identity are increasingly constructed and experienced through digital means (Table 10.1).

Digital Communities and Sport

Digital technologies have significantly reshaped how individuals interact with sport, fostering the formation of diverse digital communities that exhibit unique features and operational dynamics (Ströbel et al., 2021). This section examines various digital communities within the sport context that are believed to represent different categories of the community, but not exclusively. It discusses each community's defining characteristics, modes of operation, and illustrative real-world cases. Specifically, the section explores digital communities such as esports communities, fantasy football networks, running app groups, virtual fitness, and wearable technology platforms, VR and AR sport experiences, influencer-led social media communities, dedicated football club fan groups, and specialized Facebook groups. These illustrative digital communities can be organized into five categories (see Table 10.2) based on their primary engagement focus and operational dynamics.

TABLE 10.1 Conceptual variations of "community" in digital sport contexts

Community Concept	*Core Focus*	*Binding Factor*	*Typical Setting*	*Example*
Digital Community	Technology-mediated interaction	Digital tools and platforms	Apps, social media, AR/VR spaces	*Virtual fitness groups using wearables*
Virtual Community	Immersive digital environments	Engagement in fully virtual spaces	Virtual worlds, games, and Metaverse platforms	*FIFA eSports leagues*
Online Community	Common interests or goals	Shared topics or hobbies	Forums, websites, messaging platforms	*Reddit subreddits* on fantasy football
Brand Community	Identification with a specific brand	Shared loyalty to a brand	Online/offline hybrid	*Nike Run Club*
Social Media Community	Interaction via social platforms	Content, hashtags, influencers	Facebook, Instagram, TikTok, X (Twitter)	*Followers of a sports influencer*
Fan Community	Emotional connection to a sports entity	Passion for a team, athlete, or sport	Online, offline, and cross-platform	*Arsenal FC fans* online and at match screenings

Firstly, *digital sport gaming and interactive communities* encompass esports and fantasy sport networks, characterized by interactive gameplay and competitive engagement. Secondly, *health, fitness, and performance-tracking communities* include running apps and virtual fitness platforms, emphasizing health, wellness, and performance-oriented interactions. Thirdly, *immersive and experiential digital sport communities* consist of VR and AR platforms that provide users with enhanced, immersive sport experiences. Fourthly, *social and influencer-based digital communities*, including social media influencers and Facebook groups, facilitate personalized content and socially driven interactions. Lastly, *team affiliation and fan identity communities* are exemplified by football club fan groups characterized by a collective identity and emotional connections to specific teams. The section aims to provide a comprehensive understanding of how digital platforms have influenced fan engagement, identity formation, community building, and sport consumption in the contemporary digital era.

TABLE 10.2 Digital communities in sport

Digital Community Category	*Digital Community Type*	*Primary Engagement Focus*	*Operational Dynamics*
Digital sport gaming and interactive communities	Esports, Fantasy Sports Networks	Interactive gameplay, competitive engagement	Real-time interaction, structured games
Health, fitness, and performance-tracking communities	Running Apps, Virtual Fitness Platforms	Health, wellness, and performance-oriented interactions	Personalized tracking, goal setting
Immersive and experiential digital sport communities	Virtual Reality (VR) and Augmented Reality (AR) platforms	Enhanced immersive sport experiences	Sensory-rich environments, interactive simulations
Social and influencer-based digital communities	Social Media Influencers, Facebook Groups	Personalized content, socially driven interactions	Influencer-driven content, community engagement
Team affiliation and fan identity communities	Football Club Fan Groups	Collective identity, emotional connections	Team loyalty and shared rituals and traditions

Digital Sport Gaming and Interactive Communities

These communities focus on digital interactive gaming experiences, fostering competition, active participation, and user engagement through gameplay. This section discusses two digital community types: Esports Communities and Fantasy Sport Communities.

Esports and Digital Community

Esports, defined as organized competitive gaming involving professional players and structured events, has emerged as a rapidly growing sector within the global sport landscape, facilitated significantly by advancements in digital technologies and online communication platforms (Finch et al., 2019; Kramer et al., 2021). Central to the appeal and exponential growth of esports is the formation and sustainability of digital communities, which are groups of individuals connected through digital platforms who share common interests, activities, and collective identities related to competitive gaming experiences

(Barney & Pennington, 2023). These digital communities not only enhance fan engagement through interactive and personalized content delivery but also foster a strong sense of belonging, loyalty, and identification among members (Pizzo et al., 2022).

The League of Legends (LoL) community on Reddit, specifically the r/leagueoflegends subreddit, exemplifies the formation of digital communities within the esports landscape. This online forum serves as a central hub where millions of fans globally converge to discuss game strategies, tournament events, and player performances, thereby fostering a collective identity and sense of belonging among members (Reddit, 2025). Research underscores the significance of such digital platforms in cultivating community engagement and identity formation. For instance, Barney and Pennington (2023) highlight how online communities such as r/leagueoflegends facilitate participatory cultures, enabling fans to contribute to discussions, share content, and co-create narratives around games. This participatory engagement not only enhances fans' experience but also solidifies communal bonds within the esports ecosystem (Finch et al., 2019). Additionally, Discord, a platform popular among gamers (Robinson, 2023), hosts numerous esports communities, such as the Overwatch 2 community, the largest such community on Discord. This platform offers real-time voice and text channels that facilitate continuous communication, strategy sharing, and interaction between players and fans, effectively nurturing community cohesion (Discord, 2025).

These examples illustrate the diverse nature of digital esports communities, emphasizing their critical role in shaping contemporary sport engagement, digital fandom, and collective experiences in competitive gaming. They serve as critical facilitators of social engagement, identity construction, and fan participation within virtual competitive environments. By providing interactive spaces such as forums, social media platforms, and real-time communication channels, these digital communities have significantly amplified fan engagement, promoting sustained interaction, emotional investment, and global connectivity among esports participants (Barney & Pennington, 2023; Finch et al., 2019). Platforms such as Reddit and Discord demonstrate how these communities foster collective identity and participatory cultures, enabling fans to co-create narratives and share strategic insights, thereby strengthening communal bonds and collective experiences (Robinson, 2023).

Fantasy Sport and Digital Community

Fantasy sport is a type of online game play that allows users to assemble a hypothetical or imaginary team composed of real-life sport players (e.g., footballers) and play against other users for a specified time, whether a single day or an entire season (Abeza et al., 2024). In other words, participation in fantasy sports (e.g., FanDuel and DraftKings) involves picking players, creating a virtual team, and participating in games. Users' performance mimics the performance of real players in real games. According to Abeza et al. (2024), fantasy sports has evolved into a significant facet of contemporary sports culture today, and its growth has been compounded by advancements in digital technologies and the proliferation of online platforms, which have transformed passive sport spectators into active participants. The formation and maintenance of digital communities have been one of the key factors in the exponential growth of fantasy sports (Arizton, 2023). These communities are formed by groups of individuals connected through online platforms who share common interests and engage collaboratively (Ströbel et al., 2021) in fantasy sports activities. Communities facilitate continuous interaction, knowledge exchange, and social bonding among participants, thereby enhancing user engagement and commitment (Yadav et al., 2023).

Alongside Reddit and Discord, several other platforms host active digital communities that engage in fantasy sports on such forums as ESPN Fantasy Sports, Yahoo Fantasy Sports, CBS Sports Fantasy, FanDuel, and DraftKings. For example, The Huddle Forums demonstrates the role of digital communities in enhancing the fantasy sports experience (The Huddle, 2025). By providing an interactive platform for discussion and advice, forums such as The Huddle not only assist participants in team management but also promote a sense of community and shared identity among fantasy football enthusiasts (Ruihley & Hardin, 2011). The Huddle Forums, established in 1997, is one of the platforms dedicated to fantasy football enthusiasts (The Huddle, 2025). The forum was conceived as a gathering place for fantasy football hobbyists where they could exchange views, acquire new strategies, and gain different perspectives on players. Over the years, it has evolved into a comprehensive resource, offering an environment in which users can seek advice on draft picks, trades, starting lineups, and other team-management dilemmas (The Huddle, 2025). As Ruihley and Hardin (2011) implied, online forums such as The Huddle provide a platform for fantasy sports participants to form a collective identity among members and to engage in information exchange, strategy development, and social interaction, which could lead participants to develop a sense of belonging and identity.

Health, Fitness, and Performance-Tracking Communities

These communities emphasize health promotion, wellness, and physical fitness by integrating performance-tracking technologies and fostering user interactions focused on well-being. This section explores two specific digital community types: Running App Communities and Wearable Technology Communities.

Running Apps and Digital Community

Running applications (apps) have emerged as significant facilitators of digital communities within contemporary fitness contexts. The apps provide users with platforms to track physical activity, engage socially, and enhance exercise motivation through interactive features (Rodrigues et al., 2021). These apps integrate diverse functionalities such as performance tracking, personalized coaching, goal-setting, and gamified elements, which collectively promote sustained user engagement and encourage active participation (Yang & Koenigstorfer, 2021). A critical aspect underpinning the success of running apps is their ability to foster digital communities, enabling users to connect with like-minded peers, participate in shared challenges, and celebrate collective achievements, thus enhancing motivation and adherence to fitness routines (Stragier et al., 2016). Hence, running apps highlight how digital technologies can effectively nurture community-driven fitness cultures, promoting not only physical health but also social connectedness among users.

The Nike Run Club (NRC) digital community exemplifies how mobile applications can cultivate communities centered around physical activity and brand engagement. The app integrates features such as guided runs, progress tracking, and social sharing, enabling users to connect with others, participate in challenges, and engage with the brand in meaningful ways (NRC, 2025). By fostering a sense of community and providing personalized experiences, NRC enhances user motivation and loyalty. The app's design incorporates gamification elements—such as achievement badges, leaderboards, and milestone celebrations—that enhance user engagement and motivation (NRC, 2025). These features leverage intrinsic motivators such as competition and social recognition, encouraging consistent participation and fostering a sense of accomplishment among users (Rodrigues et al., 2021). NRC emphasizes social connectivity by enabling users to share their running achievements, participate in group challenges, and interact within a global community of runners. This social integration not only enriches the user experience but also contributes to the formation of a supportive and motivating digital community (Rodrigues et al., 2021). Adidas Running, previously known as Runtastic, represents another platform developed around digital running communities. The platform provides users with personalized training programs, social networking functionalities, and motivational challenges designed to promote ongoing user engagement (AdidasRunners, 2025). By emphasizing community building, the application enables runners to publicly share their progress, participate in virtual competitions, and actively interact with fellow community members, thereby enhancing motivation and fostering a sense of collective identity among users.

Wearable Technology and Digital Community

Virtual fitness groups represent a prominent form of digital community within the sport context (Davies et al., 2024), characterized by individuals collaboratively engaging in health and fitness pursuits through the integration of wearable technologies (Zhu et al., 2017). These groups leverage devices such as fitness trackers and smartwatches to monitor physical activity, enabling users to share progress, set collective goals, and provide mutual support (Kim, 2022). The integration of social features in fitness applications fosters a sense of community and accountability among participants. For instance, platforms such as Fitbit (Fitbit Community, 2025) and Strava (Strava Support, 2025) allow users to join virtual groups, participate in challenges, and celebrate achievements collectively, enhancing motivation and adherence to fitness routines.

Recent studies have highlighted the pivotal role of social features embedded within wearable fitness technologies in fostering community engagement and promoting sustained physical activity. For example, Zhu et al. (2017) found that the integration of social networking functionalities in wearable devices, such as sharing workout data and participating in group challenges, significantly enhances users' motivation and adherence to fitness goals. Similarly, a study by Kim (2022) reported that users who engaged with the social components of fitness apps reported higher levels of physical activity and greater satisfaction with their fitness routines. These findings underscore the importance of social connectivity in virtual fitness communities, suggesting that the communal aspects facilitated by wearable technologies are instrumental in promoting healthy behaviors and sustaining user engagement.

Digital community groups based on Fitbit and Strava platforms can serve as real-world examples to illustrate how wearable technologies facilitate collective fitness activities and community engagement. For instance, Strava supports numerous localized and global community groups, exemplified by the 3CITY Strava Challenge, a cycling-focused community in Poland, which encourages group participation in local cycling activities and strengthens social connections among its members (Strava Support, 2025). Additionally, Strava's platform hosts various Global Clubs, allowing users worldwide to join groups based on shared geographic locations, athletic disciplines, or specific interests, thus fostering international athlete connectivity and engagement (Strava Support, 2025).

Immersive and Experiential Digital Sport Communities

These communities leverage immersive digital technologies to provide interactive and experiential fan engagement within sport contexts. Two illustrative digital community types discussed in this section include VR Communities and AR Communities.

Virtual Reality and Digital Communities

VR is the computer-generated simulation of a three-dimensional image or environment with which a person can interact in a seemingly real or physical way using special electronic equipment, such as a helmet with a screen inside or gloves fitted with sensors (Abeza et al., 2024). According to the authors, VR describes the process of constructing an entirely new digital reality. One of the more popular examples is a headset that simulates situations in which the user is totally immersed. VR platforms enable the creation of digital communities in which users can experience sports events in immersive environments, fostering a sense of presence and shared experiences. VR broadcasts of games allow fans to "attend" matches together in virtual stadiums, enhancing the communal aspect of sports viewership. These technologies offer new avenues for fan engagement and community building in the digital age.

VR platforms have increasingly enabled the formation of digital communities by offering immersive social environments in which users can interact, participate in shared events, and collectively explore virtual spaces. One example is VRChat, a platform designed to facilitate unique social interactions by allowing users to fully embody custom avatars, engage in interactive social games, and connect with peers through innovative virtual experiences (VRChat, 2025). It is worth noting that even though social VR is often associated with head-mounted displays, users can engage with desktop applications, such as on VRChat, one of the oldest and most popular free-to-play social VR applications (Rzeszewski & Evans, 2024). VRChat hosts a large global community, encompassing individuals with diverse interests and backgrounds, such as gamers, musicians, creators, artists, and entertainers, who collaborate and interact in virtual environments (VRChat, 2025).

Studies (e.g., Rzeszewski & Evans, 2024) suggest that social VR platforms, such as VRChat, function as socio-technological assemblages, enabling meaningful interactions through immersive experiences, embodiment, and presence. Spatial practices within these virtual environments contribute significantly to community building, as users actively shape their spatial surroundings and establish normative social behaviors, fostering feelings of mobility, social presence, and co-presence (Rzeszewski & Evans, 2024). Although VRChat does not focus solely on sport, its immersive capabilities allow sport enthusiasts to participate in sport-related virtual activities, such as informal competitions, watch parties, and other events, highlighting its versatility in community formation around shared interests (VRChat, 2025).

Augmented Reality and Digital Community

AR is a computer technology that superimposes digital images or data on pre-existing environments (Abeza et al., 2024). While VR immerses a person into an entirely digitally created virtual environment, AR supplements an existing reality by taking digitally created content and placing it in the real-world environment (Goebert, 2020). AR integrates digital information—such as graphics, videos, or audio—into real-world environments, significantly enhancing users' perceptions and interactions during live sport events (Sohail et al., 2022). AR technologies enable sports fans to access real-time data, including player statistics and interactive visualizations, directly overlaid on their view of live matches, creating a more engaging and informed viewing experience (Goebert & Greenhalgh, 2020). This immersive technology transforms passive spectatorship into active participation, deepening fan engagement and fostering stronger emotional connections with teams and athletes (Goebert, 2020). Furthermore, AR applications facilitate interactive fan experiences by enabling real-time interactions with digital elements, such as virtual voting, augmented merchandise trials, and personalized content experiences during live sporting events (Nandi, 2025).

Some case studies illustrate the transformative potential of AR in sport, notably through enhanced broadcasting, immersive stadium experiences, and interactive social media engagements. For example, AR has been effectively integrated into National Football League (NFL) broadcasts through virtual first-down lines and player statistics, significantly enhancing viewers' understanding and enjoyment of the game (Nandi, 2025). At stadiums, such as the Miami Dolphins' Hard Rock Stadium, AR technology in mobile apps has provided fans with interactive maps, menus, and virtual seat views, greatly enriching the on-site spectator experience. Furthermore, organizations such as the National Basketball Association (NBA) utilize AR-enabled mobile applications, allowing fans to interact virtually with content, thereby enhancing their connection with the sport (Goebert, 2020). On social media, AR features such as virtual face paints and athlete-focused lenses have boosted fan engagement during major international events significantly. Examples include FIFA's AR lenses on Snapchat and similar filters employed during the Olympics on Instagram and Facebook, enabling users to creatively support their favorite teams and athletes, thus amplifying their event experience and social connectivity (Nandi, 2025). Collectively, these instances demonstrate AR's versatility and substantial role in enhancing sports events through more personalized, interactive, and memorable fan experiences.

Social and Influencer-Based Digital Communities

These communities rely on personalized content, influencer-led interactions, user-generated discussions, and social networking based on shared sport-related interests. The discussion covers two specific community types: Social Media Influencer Communities and Facebook Group Communities.

Social Media Influencers and Digital Community

Social media, as a hub for fans, serves as the online "headquarters" for partisan pride in "nations" or "countries" (e.g., #BroncosCountry, #HeatNation) and as a venue in which fans reunite outside of the stadium and access content to form bonds in the context of teams to fans, fans to athletes, and fans to fans (Abeza et al., 2017). Fans are particularly drawn to sports social media influencers, such as sports personalities with social media accounts or influencers on social media. According to Hull and Abeza (2021), social media influencers are those who have a sizeable number of followers, who respond to the needs and interests of their digital community, and whose followers consider them both sources and guides. As such, social media influencers have persuasive power and likely shape their followers' attitudes. Social media platforms have enabled the creation of digital communities centered around sport social media influencers, providing spaces in which followers can engage with specialized content, exchange experiences, and form relationships grounded in common interests and admiration for the influencer (Abeza et al., 2017). These digital communities are often anchored in the influencer's personal brand, which embodies their athletic achievements, personality traits, lifestyle choices, and values, thereby creating an emotional and aspirational bond with their followers (Hull & Abeza, 2021). According to the authors, such bonds, cultivated through consistent online interaction, facilitate deeper fan engagement and foster loyalty, reinforcing the influencer's centrality in these communities.

Digital communities centered around prominent sport influencers enable fans to actively engage in discussions, campaigns, and collective interactions (Abeza et al., 2017). For instance, soccer star Cristiano Ronaldo has cultivated one of the largest digital communities globally, with over 600 million Instagram followers (Instagram, 2025a). Ronaldo's platforms serve as interactive hubs where fans access content created on his platforms and discuss performances, fitness routines, and brand interactions, thereby reinforcing fan loyalty and enhancing community cohesion (Facebook, 2025a; Instagram, 2025a). Similarly, NBA athlete LeBron James has developed a vibrant community around his branded hashtag #StriveForGreatness. In this community, followers interact with each other through motivational content, social advocacy, and personal milestones. This engagement promotes solidarity, collective identity, and consistent fan participation across digital platforms (Instagram, 2025b).

Facebook Groups and Digital Communities

Facebook groups function as digital communities by providing several interactive features designed to facilitate engagement, connection, and collective action among members according to their athletic interests (e.g., Bodybuilding

Motivation Group, 2025; WWE Monday Night Raw Facebook Group, 2025). Key features of Facebook groups include discussion threads, content-sharing capabilities, multimedia posting (photos, videos, and live-streaming), polling tools, and event-creation options, enabling organized interactions within defined community contexts. Administrative and moderation tools help group moderators ensure content relevance, manage member interactions, and sustain community standards, thus enhancing user experience and community coherence (Facebook, 2025b).

Sport-related communities on Facebook groups, such as WWE Monday Night Raw (WWE Monday Night Raw Facebook Group, 2025) and Bodybuilding Motivation (Bodybuilding Motivation Group, 2025), leverage key platform features to foster vital and cohesive fan interactions. Discussion threads, for example, enable debates on topics such as athletic performance, match outcomes, strategic decisions, and future predictions, thereby fostering informed dialogue among fans. Multimedia posting capabilities further enrich fans' experience, allowing members to share game highlights, interviews with athletes, behind-the-scenes content, and live reactions, significantly enhancing the interactive dimension of these communities (Abeza et al., 2017). Polling tools (per Facebook, 2025b) frequently serve as mechanisms for gauging collective fan opinions and sentiments relating to player transfers, managerial changes, etc. Pumper et al. (2015), who evaluated the use of a Facebook group as part of a mHealth physical activity intervention trial, reported that members found their group motivated them and served as a tangible way of monitoring their progress. Relatedly, Looyestyn et al. (2018), in their study that examined the effectiveness of a Web-based beginners' running program for adults aged 18–50 years, delivered via a Facebook group, in increasing physical activity and cardiorespiratory fitness, reported that Facebook groups led to increased physical activity and cardiorespiratory fitness.

One example is the WWE Monday Night Raw Facebook Group, which has three million members and functions as a dedicated digital community for professional wrestling enthusiasts (WWE Monday Night Raw Facebook Group, 2025). In the group, members connect over their shared interest in professional wrestling and discuss weekly wrestling episodes, share related news, and engage in spirited debates regarding storylines, performances, and industry developments, which potentially reinforce collective identity and fan engagement. Similarly, the Bodybuilding Motivation Group, with over one million members worldwide, is an influential digital community dedicated to fitness and bodybuilding enthusiasts (Bodybuilding Motivation Group, 2025). In this group, members regularly exchange workout routines, nutritional advice, and motivational content, thereby promoting mutual support, empowerment, and a collective pursuit of fitness-related goals (Bodybuilding Motivation Group, 2025).

Team Affiliation and Fan Identity Communities

This community type centers specifically on collective passion, loyalty, emotional engagement, and shared identity linked to supporting specific football clubs. This section explores Football Club Fan Communities.

Football Club Fans and Digital Communities

Online football club fan communities have proliferated on platforms such as forums, blogs, bulletin boards, and message boards where supporters engage, discuss club activities, and share team-related content. Members of such online communities usually comprise individuals who seek to share common goals, interests, and/or needs with fellow members (Kim & Kim, 2017). In contrast to social media platforms, which constitute open environments accessible to both supporters and nonsupporters, online fan communities thrive on digital spaces reserved primarily for dedicated fans (Kim & Kim, 2017). As the authors noted, official fan communities frequently require members to pay an annual subscription fee; however, even when no fee is charged, platform administrators retain the authority to remove antagonistic participants, thereby maintaining the integrity and exclusivity of the fan community.

Online football fan communities transcend geographical constraints by offering access to fans worldwide (Abeza et al., 2017). Members can use digital platforms as spaces to connect, communicate, and demonstrate allegiance to their respective teams. These digital communities facilitate continuous interactions among fans, enabling them to actively participate in discussions related to team performance, player transfers, and other club-related developments (Kim & Kim, 2017). Such ongoing interactions foster a collective identity and a shared sense of belonging, enhancing both emotional attachment and loyalty toward the club (Abeza et al., 2017). These digital communities not only share match highlights or general club news but also analyze team strategies, evaluate managerial decisions, and express their passion through media-sharing practices, interactive events, and fan-driven content creation such as blogs, podcasts, and livestream commentaries (Barcaforum, 2025; RedCafe, 2025).

One example of an online football club fan community is the FC Barcelona Fans Forum (Barcaforum, 2025), which serves as a centralized hub for FC Barcelona supporters worldwide. The hub facilitates discussions on various topics such as match analyses, player performances, transfer speculations, and club-related news. The forum's (see Barcaforum, 2025) structure includes sections such as "General FC Barcelona Talk," "FC Barcelona News," and "Match Chat," enabling organized and in-depth conversations among fans. The platform provides a space for continuous engagement among members, which fosters a sense of community and shared identity, contributing to the global fan culture surrounding the club. Another example of an online football club fan community is the RedCafe forum, one of Manchester United's forums, which serves as a central hub for supporters worldwide, facilitating discussions on various topics such as match analyses, player performances, transfer speculations, and club-related news (RedCafe, 2025). The forum's structure includes sections such as "Manchester United Forum," "Player Performances Forum," and "Match Day" (RedCafe, 2025), all of which serve as spaces for continuous engagement and shared identity.

Practice in Action

Reddit's Subreddits Dedicated to Sport

The digital platform Reddit hosts numerous sport-focused subreddits that serve as conversational forums for fan interactions and discourse. Notable examples such as r/NBA, and r/NFL function as interactive spaces in which participants share analytical insights, news updates, and real-time event discussions. These subreddits have evolved into influential platforms significantly shaped by user-generated content, and thus serve as collective narratives within digital sport communities.

Several factors drive engagement within these digital sport communities. Firstly, free participation, facilitated by Reddit's upvoting mechanism (i.e., approval of or agreement with a post by means of a particular icon), encourages community members to make worthy contributions, thereby promoting informed debates and collective validation. Secondly, Reddit's inherent anonymity and demographic diversity allow contributors to engage in candid, open discourse, fostering valuable interactions among diverse participant groups. Thirdly, the asynchronous nature of Reddit's threaded discussions supports reflective conversations, going beyond the temporal constraints typical of live sport events. Additionally, community moderation plays a crucial role in sustaining respectful and constructive dialogue; volunteer moderators and automated moderation systems collaborate to enforce community guidelines, ensuring decorum and appropriate user conduct.

This case illustrates how technologically driven content curation, combined with participatory engagement structures, can create inclusive environments in which both casual fans and dedicated sport enthusiasts are encouraged to actively contribute to the ongoing collective narrative, thus enhancing overall community enthusiasm.

Trends and Current Issues

One of the trends and current issues in digital communities in sport is the Metaverse and digital convergence. The emergence of the Metaverse implies an integrated digital ecosystem merging virtual and real-world experiences within sport contexts, influencing digital sport communities. Interactive virtual spaces, such as digital stadiums and fan zones, foster deeper fan interaction, allowing digital communities to congregate in persistent virtual environments. Relatedly, there is an integration of immersive technologies into daily fan engagement. Immersive technologies such as AR, VR, and the Metaverse increasingly embed into routine digital community interactions. Digital sport communities benefit from fully interactive virtual events, including live games and virtual meet-and-greets, enhancing social connectivity and community cohesiveness. Enhanced digital storytelling provided by these immersive experiences offers richer narratives, deepening emotional ties within communities and promoting sustained community participation. Similarly, enhanced personalization and interactivity are the other trends and current issues in digital communities in sport. Digital sport communities increasingly benefit from personalized fan experiences enabled by artificial intelligence (AI) and machine learning. Advanced algorithms analyze community behavior to deliver tailored content, enhancing member satisfaction and engagement. Real-time interactive tools, including live polls and virtual Q&A sessions, strengthen community interaction, transforming digital platforms into highly participative spaces.

Summary

This chapter examined the emergence and development of digital communities in sport, highlighting the shift from historically localized, face-to-face fan interactions to dynamic digital communities. Historically, sport fandom was localized, characterized by face-to-face interactions and shared cultural practices. However, the proliferation of digital platforms has facilitated the creation of vibrant online sport communities, contributing to enhanced fan connectivity, interactivity, and global participation. Digital communities have thus redefined traditional sport fandom, transforming it into an interactive, globally accessible phenomenon that transcends geographical boundaries and expands opportunities for social engagement and identity formation.

Key Resources

- **Front Office Sports**: https://frontofficesports.com/
- **Nielsen Sports**: https://www.nielsen.com/insights/sports/
- **The Conversation**: https://theconversation.com/

References

Abeza, G., & King-White, R. (2024). *Sport and Social Media in Business and Society*. Taylor & Francis.

Abeza, G., O'Reilly, N., & Seguin, B. (2019). Social media in relationship marketing: The perspective of professional sport managers in the MLB, NBA, NFL, and NHL. *Communication & Sport*, *7*(1), 80–109.

Abeza, G., O'Reilly, N., Seguin, B., & Nzindukiyimana, O. (2017). Social media as a relationship marketing tool in professional sport: A netnographical exploration. *International Journal of Sport Communication*, *10*(3), 325–358.
Abeza, G., O'Reilly, N., Séguin, B., & Nzindukiyimana, O. (2017). The world's highest-paid athletes, product endorsement, and Twitter. *Sport, Business and Management: An International Journal*, *7*(3), 332–355.
Abeza, G., Wanless, E. & Mamo, Y. (2024). Emerging developments in sport management. In P. Pedersen & L. Thibault (Eds.), *Contemporary Sport Management*. HK.
AdidasRunners. (2025). Welcome to adidas runners. Retrieved from https://www.adidas.com/us/adidasrunners/
Arizton. (2023). Fantasy sports market: Global outlook and forecast 2019–2024. Retrieved from https://www.arizton.com/news/press-release/global-fantasy-sports-industry-2024
Barcaforum. (2025). FC Barcelona forum. Retrieved from https://www.barcaforum.com/
Barney, J., & Pennington, N. (2023). An exploration of esports fan identity, engagement practices, and motives. *Telematics and Informatics Reports*, *11*, 100081.
Bodybuilding Motivation Group. (2025). *Facebook Groups*. Retrieved May 2, 2025, from https://www.facebook.com/groups/bodybuildingmotivationgroup/
Boyd, D. M., & Ellison, N. B. (2007). Social network sites: Definition, history, and scholarship. *Journal of Computer-Mediated Communication*, *13*(1), 210–230.
Constantin, V. D., Platon, O., & Orzan, G. (2014). Brand community formation: A critical review. *Annales Universitatis Apulensis Series Oeconomica*, *16*(2), 123–131.
Davies, M., Hungenberg, E., Aicher, T. J., & Newland, B. L. (2024). Work [out] from home: Examining brand community among connected fitness brand users. *International Journal of Sport Management and Marketing*, *24*(2), 113–136.
Discord. (2025). Overwatch 2. Retrieved from https://discord.com/servers/overwatch-2-94882524378968064
Dover, Y., & Kelman, G. (2018). Emergence of online communities: Empirical evidence and theory. *PloS One*, *13*(11), e0205167.
Facebook. (2025a). Cristiano Ronaldo official Facebook page. Retrieved May 2, 2025, from https://www.facebook.com/Cristiano/
Facebook. (2025b). Groups. Retrieved from https://www.facebook.com/help/groups
Finch, D. J., O'Reilly, N., Abeza, G., Clark, B., & Legg, D. (2019). *Implications and Impacts of eSports on Business and Society: Emerging Research and Opportunities*. IGI Global.
Fitbit Community. (2025). Community. Retrieved from https://community.fitbit.com/t5/Community/ct-p/EN
Gantz, W., & Lewis, N. (2014). Sports on traditional and newer digital media: Is there really a fight for fans? *Television & New Media*, *15*(8), 760–768.
Gibbons, T. (2019). Arsenal Fan TV, football fandom and banter in the new media era. *Soccer & Society*, *20*(4), 558–574.
Glebova, E., Su, Y., Schut, P. O., & Desbordes, M. (2025) Emerging digital technologies as a game changer in the sport industry. *Frontiers in Sports and Active Living*, *7*, 1605138.
Goebert, C. (2020). Augmented reality in sport marketing: Uses and directions. *Sports Innovation Journal*, *1*, 134–151.
Goebert, C., & Greenhalgh, G. P. (2020). A new reality: Fan perceptions of augmented reality readiness in sports marketing. *Computers in Human Behavior*, *106*, 106231.
Hamari, J., & Sjöblom, M. (2017). What is eSports and why do people watch it? *Internet Research*, *27*(2), 211–232.
Hammond, M. (2017). What is an online community? A new definition based around commitment, connection, reciprocity, interaction, agency, and consequences. *International Journal of Web Based Communities*, *13*(1), 118–136.

Hull, K., & Abeza, G. (2021). Introduction to social media in sport. In G. Abeza, N. O'Reilly, J. Sanderson & E. Fredrick (Eds.), *Social Media in Sport: Theory and Practice* (pp. 1–28). World Scientific.

Hutchins, B., & Rowe, D. (2012). *Sport Beyond Television: The Internet, Digital Media and the Rise of Networked Media Sport*. Routledge.

Instagram. (2025a). Cristiano Ronaldo official Instagram account. Retrieved May 2, 2025, from https://www.instagram.com/cristiano/

Instagram. (2025b). LeBron James official Instagram account. Retrieved May 2, 2025, from https://www.instagram.com/kingjames/

Kian, E. M., Burden Jr, J. W., & Shaw, S. D. (2011). Internet sport bloggers: Who are these people and where do they come from? *Journal of Applied Sport Management*, *3*(1), 13.

Kim, H. M. (2022). Social comparison of fitness social media postings by fitness app users. *Computers in Human Behavior*, *131*, 107204.

Kim, M. S., & Kim, H. M. (2017). The effect of online fan community attributes on the loyalty and cooperation of fan community members: The moderating role of connect hours. *Computers in Human Behavior*, *68*, 232–243.

Kramer, K., Wagner, D., & Scheck, B. (2021). Reaping the digital dividend? Sport marketing's move into esports: Insights from Germany. *European Journal of International Management*, *15*(2/3), 339–366. https://doi.org/10.1504/EJIM.2021.10032384

Looyestyn, J., Kernot, J., Boshoff, K., & Maher, C. (2018). A web-based, social networking beginners' running intervention for adults aged 18 to 50 years delivered via a Facebook group: Randomized controlled trial. *Journal of Medical Internet Research*, *20*(2), e67.

Muniz Jr, A. M., & O'Guinn, T. C. (2001). Brand community. *Journal of Consumer Research*, *27*(4), 412–432.

Nandi, S. (2025). Augmented reality in sports: Enhancing fan engagement. *AZoAi*. Retrieved from https://www.azoai.com/article/Augmented-Reality-in-Sports-Enhancing-Fan-Engagement.aspx

NRC. (2025). Nike running club. Retrieved from https://www.nike.com/nrc-app

Ovadia, S. (2015). More than just cat pictures: Reddit as a curated news source. *Behavioral & Social Sciences Librarian*, *34*(1), 37–40.

Pizzo, A. D., Su, Y., Scholz, T., Baker, B. J., Hamari, J., & Ndanga, L. (2022). Esports scholarship review: Synthesis, contributions, and future research. *Journal of Sport Management*, *36*(3), 228–239.

Porter, C. E. (2004). A typology of virtual communities: A multi-disciplinary foundation for future research. *Journal of Computer-Mediated Communication*, *10*(1), JCMC1011.

Pumper, M. A., Mendoza, J. A., Arseniev-Koehler, A., Holm, M., Waite, A., & Moreno, M. A. (2015). Using a Facebook group as an adjunct to a pilot mHealth physical activity intervention: A mixed methods approach. *Annual Review of Cybertherapy and Telemedicine*, *2015*, 97–101.

Rauschnabel, P. A. (2021). Augmented reality is eating the real-world! The substitution of physical products by holograms. *International Journal of Information Management*, *57*, 102279.

RedCafe. (2025). Home. Retrieved from https://www.redcafe.net/

Reddit. (2025). r/leagueoflegends. Retrieved from https://www.reddit.com/r/leagueoflegends/?rdt=45773

Robinson, B. (2023). Governance on, with, behind, and beyond the Discord platform: A study of platform practices in an informal learning context. *Learning, Media and Technology*, *48*(1), 81–94.

Rodrigues, I., Lopes, J. M., Borges, A., Oliveira, J., & Oliveira, M. (2021). How can Gamified applications drive engagement and brand attitude? The case of Nike run club application. *Administrative Sciences*, *11*(3), 92.

Rowe, D. (2011). *Global Media Sport: Flows, Forms and Futures*. Bloomsbury Academic.
Ruihley, B. J., & Hardin, R. L. (2011). Message boards and the fantasy sport experience. *International Journal of Sport Communication*, *4*(2), 233–252.
Rzeszewski, M., & Evans, L. (2024). Social relations and spatiality in VR-Making spaces meaningful in VRChat. *Emotion, Space and Society*, *53*, 101038.
Sanderson, J., & Kassing, J. W. (2011). Tweets and blogs: Transformative, adversarial, and integrative developments in sports media. In A. C. Billings (Ed.), *Sports Media: Transformation, Integration, Consumption* (pp. 114–127). Routledge.
Sohail, Z., Firdos, A., Ikram, S., & Talha, M. (2022). The impact of virtual reality and augmented reality on sport psychology. *Revista de Psicología del Deporte (Journal of Sport Psychology)*, *31*(1), 217–226.
Stewart, B., Smith, A., & Nicholson, M. (2003). Sport consumer typologies: A critical review. *Sport Marketing Quarterly*, *12*(4), 206–216.
Stragier, J., Vanden Abeele, M., Mechant, P., & De Marez, L. (2016). Understanding persistence in the use of online fitness communities: Comparing novice and experienced users. *Computers in Human Behavior*, *64*, 34–42.
Strava Support. (2025). Clubs on Strava. *Strava Support*. Retrieved May 1, 2025, from https://support.strava.com/hc/en-us/articles/216918347-Clubs-on-Strava
Ströbel, T., Woratschek, H., & Horbel, C. (2021). Guest editorial – Digital transformation in sport: The disruptive potential of digitalization for sport management research. *European Sport Management Quarterly*, *21*(1), 1–12.
The Huddle. (2025). *About the Huddle*. Retrieved May 1, 2025, from https://thehuddle.com/about-2/
VRChat. (2025). *Community Guidelines and Features*. Retrieved May 1, 2025, from https://hello.vrchat.com
Wagner, D. (2021). Online communities in sport. In G. Abeza, N. O'Reilly, J. Sanderson, & E. Frederick (Eds.), *Social Media in Sport* (pp. 57–81). World Scientific. https://doi.org/10.1142/9789811237669_0003
Wagner, D. (2023). A review and research Agenda for brand communities in sports. *International Journal of Sport Communication*, *16*(3), 335–343. https://doi.org/10.1123/ijsc.2023-0114
Wann, D. L., Melnick, M. J., Russell, G. W., & Pease, D. G. (2001). *Sport Fans: The Psychology and Social Impact of Spectators*. Routledge.
WWE Monday Night Raw Facebook Group. (2025). *Facebook Groups*. Retrieved May 2, 2025, from https://www.facebook.com/groups/301306270260958/
Yadav, J., Yadav, R., Sahore, N., & Mendiratta, A. (2023). Digital social engagements and knowledge sharing among sports fans: Role of interaction, identification, and interface. *Technological Forecasting and Social Change*, *195*, 122792.
Yadav, M. S., & Pavlou, P. A. (2020). Technology-enabled interactions in digital environments: A conceptual foundation for current and future research. *Journal of the Academy of Marketing Science*, *48*, 132–136.
Yang, Y., & Koenigstorfer, J. (2021). Determinants of fitness app usage and moderating impacts of education-, motivation-, and gamification-related app features on physical activity intentions: Cross-sectional survey study. *Journal of Medical Internet Research*, *23*(7), e26063.
Zhu, Y., Dailey, S. L., Kreitzberg, D., & Bernhardt, J. (2017). "Social networkout": Connecting social features of wearable fitness trackers with physical exercise. *Journal of Health Communication*, *22*(12), 974–980.

11

TECHNOLOGY IN ADAPTIVE SPORT AND RECREATION

Irina Perfilova, Robin Hardin, and Joshua R. Pate

Introduction

The broad-based term to describe sports or recreational activities that are tailored or modified for people with disabilities is adaptive sport and recreation (McCarty et al., 2025; U.S. Olympic & Paralympic Committee, n.d.). Categories of adaptive sports and recreation can be broadly classified based on the nature of the sport or activity, the type of disability they accommodate, or the modifications made (Braza et al., 2018; Iverson et al., 2021). The modifications are made so people with physical, sensory, or intellectual disabilities can participate in sport and recreation. Adaptions are typically to equipment, rules, or competition styles to allow for participation (Hardin & Pate, 2025). Adaptive sport and recreation may seem a minor aspect of the wider sport industry, but it is essential to understand its broader impact and how technology has affected participation and consumption.

Approximately 1.3 billion people worldwide experience some form of disability (World Health Organization [WHO], n.d.), while the United States Census Bureau reports that approximately 13.4% of the US population lives with a disability (United States Census Bureau, 2024). This substantial segment of the population emphasizes the need for continuous improvement in adaptive sports technology to ensure broader access and participation. As the popularity of adaptive sport and recreation continues to grow, it creates new opportunities for technological advancements and expanded participation (De Luigi, 2024; Khurana et al., 2021; Legg & Darcy, 2025). Technological innovations have allowed for increased participation opportunities and access to more sports and recreational activities through equipment adaptations and development. For instance, adaptive equipment for snowsports and cycling has made the ability to participate in those activities nearly accessible for all. Assistive technological advances have made the consumption of sports much more spectator-friendly for spectators with disabilities.

DOI: 10.4324/9781003588405-14

This chapter explores technology and innovations in adaptive sport and recreation. The chapter opens with defining adaptive sport and describing the differences between adaptive and assistive technologies. Then, a brief history of the Paralympic Movement is discussed and its connection to the development of adaptive sport and recreation. Meanwhile, technology will be outlined and divided into two categories to offer readers a deeper understanding of its role in industry: (1) technology designed for athletes and (2) technology aimed at consumers of adaptive sport and recreation. Different examples will be introduced for each category, describing technological advancements and innovations. This foundational knowledge can assist individuals with no prior experience in adaptive sport and recreation to develop a clearer understanding of how technology enhances accessibility, performance, and inclusion for athletes with disabilities.

Background

Initially, adaptive sport and recreational activities were developed as a part of the rehabilitation process for veterans from World War II, Korea, and the Vietnam War. Therefore, participation was primarily driven by rehabilitation programs when returning veterans sought ways to remain active despite their injuries (Brittain, 2014; Guttmann, 1952). In the early years of adaptive sport and recreation, technology was rudimentary and heavily focused on basic functionality rather than performance or customization. Adaptive equipment was often improvised from everyday items or modified from medical devices used for rehabilitation (Lee & Uihlein, 2019). For instance, early wheelchair sports relied on heavy, steel-frame wheelchairs that were difficult to maneuver and unsuitable for high-speed movement or sudden turns (Brady, 2023). Athletes had to adapt their playing style to the equipment's limitations, with little consideration given to sports-specific designs.

Prosthetics were similarly basic, primarily designed for walking rather than athletic performance. Van Phillips invented what is called the Flex-Foot prosthetic limb after losing his leg in a water-skiing accident in 1976 (Van Phillips, n.d.). He believed that available prosthetics at that time were not suitable for athletic movements, so he sought to create a better and sports-focused solution. His invention revolutionized the prosthetics industry by introducing carbon graphite materials, which were designed to mimic the natural energy return of a human foot. This innovation not only improved mobility for amputees but also became the foundation of the modern running blades used by elite Paralympic athletes (Hobara, 2014; Nagamatsu, 2022).

This era also saw a significant shift in the mid-late 20th century, when the primary focus moved from rehabilitation to performance. Athletes began competing at higher levels, including the Parapan American Games and the Paralympic Games. Organizations such as the International

Paralympic Committee (IPC) played an important role in promoting these technological advancements by partnering with researchers and engineers to develop sports-specific adaptive equipment. Such partnerships included the World Health Organization, Toyota Mobility Foundation, and Bridgestone (Bridgestone Americas Press Center, 2024; Toyota Mobility Foundation, 2024; World Health Organization, 2024a). The range of technologies continues to grow to enhance both participation and performance. The US Department of Veterans Affairs (VA) offers adaptive sports grants to plan, develop, manage, and implement adaptive sports programming across the country. Several nonprofit organizations in the United States, such as Challenge Aspen, Move United, Adaptive Adventures, Challenged Athletes Foundation (CAF), and Lakeshore Foundation, also offer grants to assist individuals with expenses related to equipment, training, or competition in adaptive sport and recreation. In addition, carbon-fiber frames have revolutionized sports-specific wheelchairs, making them lighter, faster, and more maneuverable, while carbon-fiber running blades have allowed athletes with lower-limb amputations to compete at the highest levels in track and field (De Luigi, 2024). These innovations have redefined participation and performance, opening new opportunities for athletes at all levels (Matsuwaka & Latzka, 2019).

Topical Exploration and Key Concepts

Broadly defined, adaptive sport and recreation refers to modified or specifically adopted organized sports that meet the needs of individuals with physical, sensory, or intellectual disabilities (Hardin & Pate, 2025). Adaptive sport and recreation provide individuals with disabilities opportunities to engage in physical activity, competitive sports, recreation, leisure activities, and community building (Jamieson & Wijesundara, 2025). These activities often require specialized equipment and technological innovations that accommodate diverse abilities while enhancing performance and accessibility.

Not all individuals involved in adaptive sport and recreation identify as "athletes." For the purpose of this chapter, the term "athlete" refers to individuals with physical or intellectual disabilities who compete in organized adaptive sports, while "participant" includes those who engage in adaptive recreational activities without a competitive focus. This distinction recognizes the diverse ways people experience and engage in adaptive sport and recreation.

Technology: Assistive vs Adaptive

Technology has become an essential component that enhances access and performance for athletes and participants of adaptive sport and recreation.

There are various technologies used in adaptive sports, including equipment such as prosthetics, sports-specific wheelchairs, and wearable devices. These advancements not only improve functionality and athletic performance but also reduce barriers to participation, which promotes greater independence and self-expression. Technology often focuses on increasing comfort and accessibility for recreational participants, while competitive athletes benefit from precision-engineered equipment designed to maximize athletic performance.

The development of technology in adaptive sports is closely tied to advances in both assistive technology and the growing recognition of the Paralympic Movement. Assistive technology is "an umbrella term for assistive products and their related systems and services" (World Health Organization, 2024a, Overview section). This definition emphasizes the wide range of people who can benefit from assistive technology, noting that "most people will need assistive technology at some point in their lives, especially as they age" (World Health Organization, 2024b, Key facts section). While the term "assistive technology" is widely used in healthcare and rehabilitation, it is less commonly applied in the context of adaptive sport and recreation. Unlike assistive technology, which often serves a functional or rehabilitative purpose, technologies in adaptive sport and recreation are designed with sports-specific goals in mind, such as maximizing speed, agility, or communication during competition.

Technology Designed for Athletes and Participants

Technology in adaptive sport and recreation varies between competitive athletics and noncompetitive participation because each serves a different purpose. For competitive athletes, technology focuses on maximizing athletic performance, as in the case of advanced prosthetics and racing wheelchairs. For recreational participants, the emphasis is on accessibility, comfort, and inclusivity. Both types of technology enhance participation, but their design and application align with the specific needs and goals of athletes or participants (Legg & Darcy, 2025).

Wheelchair sports include a variety of competitive and recreational activities where manual or powered wheelchairs are used. These sports provide opportunities for athletes and participants with spinal cord injuries, lower-limb impairments, amputations, and other mobility-related disabilities to engage in sport or recreation (Rayes et al., 2022). The most widely recognized wheelchair sports include wheelchair basketball, wheelchair rugby, and wheelchair tennis, with emerging sports like wheelchair football, wheelchair lacrosse, and wheelchair badminton.

Although wheelchair sports fall under the same category, each sport requires a specialized wheelchair designed to increase efficiency and ensure the safety of the users, while also adhering to the specific rules and regulations of the game. For example, wheelchair basketball chairs prioritize agility and quick change of directions (de Groot et al., 2003), while wheelchair rugby chairs are reinforced for high-impact collisions (Usma-Alvarez et al., 2011). Particularly, chairs for wheelchair rugby are equipped with aluminum "wings" on the front wheels to enhance agility and maneuverability for offensive players, while defensive players' wheelchairs feature a front-mounted hook designed to trap and block opponents effectively (Bakatchina et al., 2022). Additionally, while basic features of the wheelchair are the same for every player in each sport, some players might need to add customizations to their chairs based on their physical ability. The primary goal of wheelchair sports technology is to facilitate integration between the athlete and their wheelchair, meaning that these technologies ensure that the wheelchair functions as an extension of the athlete's body (Cooper & De Luigi, 2014).

Technologies in wheelchair sports include lightweight frame materials such as titanium, aluminum, and carbon fiber, which in turn enhance durability, speed, and maneuverability while reducing strain on the athlete. Customization in all forms includes seating systems to promote proper posture and pressure distribution, which minimizes the risk of injury and maximizes efficiency during play. Additionally, cambered wheels improve stability and turning ability, while high-performance push rims (Cooper & De Luigi, 2014). Emerging technology wheelchairs also incorporate shock absorption systems to reduce impact, particularly in contact sports like wheelchair rugby (Giwnewer et al., 2020; Mason et al., 2010).

Another type of wheelchairs are those designed for racing, and those with running prosthetics are commonly used in para-athletics (track and field). Racing wheelchairs, often referred to as racing chairs, are sometimes mistakenly called handcycles due to the hand-driven propulsion system used by athletes. However, unlike handcycles, which have a gear and chain system like bicycles, racing wheelchairs rely on push-rim propulsion, where athletes generate speed by pushing directly on the hand rims attached to the wheels. Racing wheelchairs are designed for maximum speed and aerodynamics and incorporate several key technologies to improve athletic performance. First, lightweight frames, typically made from carbon fiber, aluminum, or titanium, are designed to reduce weight, improve acceleration, and maximize energy efficiency. Second, three-wheel designs provide better stability and aerodynamics, with a long front wheel. Third, custom seating and steering allow athletes to adopt either a kneeling or seated position, with adjustable footplates for optimized power transfer. Steering on a racing chair is controlled through body shifts or manual levers. Finally, to maximize efficiency and protect their hands, athletes wear specialized racing gloves made from high-friction materials like rubber for better grip and force transfer. 3D-printed gloves have emerged on the market that offer more customization options for athletes and participants to maximize performance and comfort. The ability for 3D-printed gloves to be custom-fit to a participant's hand allows for a tailored glove to be made for someone whose hands may not conform to the typical anatomical structure or mobility of the hands. Someone whose hands have muscle constriction or limited dexterity can have gloves custom-fitted to their hands. 3D gloves can also augment grip strength and grip assurance.

Paratriathlon is a three-stage race consisting of swimming, cycling, and running, which requires additional technology to accommodate these different events. The competition requires athletes to quickly transition between each of the three events. Therefore, technological advancements are focused on minimizing the time spent during transitions. Key innovations include quick-release prosthetic attachments, which allow athletes to switch prosthetics fast between the cycling and running stages. For the swimming stage, advanced technology can be seen in wetsuit designs, which feature tear-away zippers and flexible neoprene materials. Adaptive handcycles and tandem bikes are used for the cycling stage. Those bikes incorporate electronic gear shifting, ergonomic hand cranks, and custom-fit seating to enhance speed and maneuverability (Gordon & De Luigi, 2020). Racing wheelchairs for the running stage are built with quick-lock seat straps to ensure smoother transitions and adaptability to road conditions. Additionally, wearable technology such as biometric sensors and analytics is widely used by athletes to track heart rate, power output, and fatigue levels for optimizing performance.

It is critical to ensure athletes' and participants' safety in water, whether it is competitive water sports or recreational activities. In adaptive water skiing, specialized equipment, such as sit-skis allows individuals with lower-limb impairments to experience water skiing. Companies design and manufacture adaptive water ski equipment tailored to various skill levels. Their products are developed based on feedback from athletes, coaches, and other stakeholders, ensuring functionality and performance align with user needs. For those interested in adaptive kayaking, innovations include specialized paddling systems that accommodate a range of physical and dexterity abilities. Manufacturers offer adaptive kayak paddles with adjustable features, which offer individuals with limited upper body strength or mobility to paddle independently. Adaptive surfing has also progressed, with organizations providing programs and equipment to empower people with disabilities.

Technology as a Tool for Inclusion and Integration

Watching sports is just as important as playing them for many people, but not everyone has equal access to that experience. In adaptive sports, where athletes with disabilities compete at elite levels, it is essential to make sure that all fans, including those with disabilities, can fully engage in watching and following the games. The Paris 2024 Paralympic Games set a new standard for inclusivity through the integration of artificial intelligence (AI) and other technological advancements. One of the most interesting innovations was the Vision Pad, a tactile tablet designed for visually impaired fans. Vision Pad allowed fans to "feel" the movement of athletes and the flow of competition through haptic (vibration-based) feedback. This device represented a major step forward in making live sporting events more inclusive for spectators with vision impairments. Another Paris 2024 innovation was AI-generated audio descriptions, which provided play-by-play commentary for visually impaired fans attending events both in-person and remotely (Leicester, 2024). Unlike traditional commentary, this system used machine learning to deliver detailed and immediate descriptions of fast-paced sporting action.

Technology in adaptive sport is not just about improving performance—it is also promoting accessible and inclusive opportunities as well as aiding in understanding the experiences of athletes in adaptive sports. In 2021, the IPC introduced a virtual reality (VR) tool named RAW: Emotions Unite Us (Bringing People Together Around the Paralympics, 2021). It is an interactive WebXR experience produced in collaboration between the IPC, Meta, and PHORIA with the purpose of positively impacting perceptions of disability by immersing users in the journeys of five paralympians. The technology provides an "immersive WebXR visual and spatial audio experience that recreates phenomenal moments by Para athletes" (Bringing People Together Around the Paralympics, 2021, para. 3). The technology provides uses the experience of being "in the moment" with the athletes (Bringing People Together around the Paralympics, 2021, para. 3). It is also important to understand the concepts that have influenced the perception of adaptive sports and recreation: the medical model and the social model.

The medical model views disability as a problem that exists within a person – something that needs to be fixed or treated (McCarty et al., 2025). In contrast, the social model of disability suggests that the issue is not the disability itself, but the barriers in society that make it difficult for people with disabilities to get involved in sports and recreation (Haegele & McKay, 2025). In this model, technology helps break down these barriers by creating more inclusive environments. As one example, blind cap technology was first introduced at the Rio 2016 Paralympics and appeared in Paris 2024 with further improved functionality (Samsung Spain Launches Smart Cap to Help Blind Swimmers Compete, 2016). In particular, the blind cap provides vibrational alerts to swimmers when they are nearing the pool wall, which prevents them from hitting the wall and allows precise flip-turns. Tools such as guidance systems for blind athletes or smart stadiums with accessible features allow people with disabilities to participate in sports in a way that is just as engaging and fair as for anyone else. The Paralympic program continues to expand with new sports (e.g., para climbing, para surfing), and this expansion will require advanced adaptive technologies to accommodate the demands of new disciplines (LA28 Paralympic Sports Starting Lineup, n.d.).

Trends and Current Issues

One of the fastest-growing adaptive sports in terms of technological advancements is aMTB, which stands for Adaptive Mountain Biking. This sport has gained popularity due to improved bike designs, safety features, and increased accessibility for riders with various disabilities (Broudy, 2023; Bruce, 2023). Adaptive mountain bikes (aMTBs) come in several forms, including three-wheeled recumbent bikes, four-wheeled downhill rigs, and electric-assisted handcycles. These adapted mountain bikes incorporate full-suspension frames, wider tires for stability, and electric-assist motors, which is a new opportunity for extreme sports lovers. Those features allow riders to tackle rough terrain, steep inclines, and technical descents with greater control. In an interview, adaptive mountain biker Annijke Wade described aMTB as a transformative journey that has been both challenging and rewarding. After a life-changing accident in July 2021 left her paralyzed from the chest down, Wade was determined to return to the trails (Kazunas, 2024). She immersed herself in researching adaptive mountain biking, selecting a specialized three-wheeled bike that allowed her to navigate off-road terrains. Wade emphasizes the importance of patience and self-acceptance, noting that while the equipment and her body have changed, her passion for mountain biking remains unwavering. She encourages others to embrace the learning process and explore adaptive outdoor activities (Kumar, 2023).

One issue that is persistent, despite the growth and technological advancements in adaptive sports and recreation, is the lack of awareness and visibility (Pate et al., 2023). This issue affects both able-bodied and disabled populations (Braza et al., 2018; Iverson et al., 2021). Many able-bodied individuals are unaware that adaptive sports exist as competitive and recreational opportunities. Simultaneously, many individuals with disabilities may not be aware of the sports programs available to them. Duvall et al. (2021) noted a significant gap in the availability of sports programs for individuals who use power wheelchairs. Similarly, Jamieson and Wijesundara (2025) acknowledged innovative technologies but also pointed out that "there is still a significant part of this population that does not participate in any form of exercise" (p. 43). The lack of awareness and availability of programs has been a persistent barrier for participation in adaptive sports and recreation (McCarty et al., 2025).

Another significant issue in adaptive sports and recreation is the high cost of specialized equipment, which can make participation difficult for many athletes with disabilities. Unlike standard sports gear, adaptive sports equipment is custom-built, designed to accommodate specific disabilities and optimize performance (Jamieson & Wijesundara, 2025). For example, gloves for competitive wheelchair racing start are approximately $300 US, and high-performance racing wheelchairs can cost upwards of $10,000 US. Similarly, advanced running prosthetics, such as carbon fiber blades, range from $5,000 US to more than $50,000 US. These costs create a financial barrier that prevents many athletes from accessing the technology necessary to train and compete at a high level.

In addition to cost issues, access to adaptive sports equipment and facilities remains a major challenge. Many athletes with disabilities struggle to find local programs, training centers, or coaches who are knowledgeable and competent in adaptive sports. In some regions, there is limited access to specialized wheelchairs, prosthetics, and training technology, making it difficult for athletes to develop their skills. Additionally, many sporting venues lack accessibility features, such as ramps, adaptive gym equipment, or specialized surfaces, further restricting opportunities for participation (Moorman, 2025; Paoli & Robles, 2025; Stran & Hardin, 2025).

Questions about fairness in competition have arisen as well as technology in prosthetics and wheelchairs has evolved (DeVillez et al., 2022). Innovations in sports-specific equipment promote greater performance; however, these advancements have led to debates about whether some technologies provide an unfair competitive advantage. Some prosthetic limbs may offer greater energy return than biological limbs, leading to discussions about regulation in Paralympic sports (De Luigi, 2024).

Practice in Action

Challenge Aspen, a nonprofit organization based in Snowmass Village, Colorado, is dedicated to providing year-round outdoor adaptive sport and recreational opportunities for participants with physical or intellectual disabilities. Programming includes adaptive skiing, adaptive snowboarding, mountain biking, archery, paddleboarding, and a plethora of other activities. They offer programs to a wide range of participants, from youth to high school students to adults as well as programs specifically for veterans. A primary goal of Challenge Aspen is to negotiate barriers for participants with disabilities to participate in these activities. Technology and innovation have played a vital role in navigating these barriers, and that is certainly evident in adaptive snowsports.

Adaptive skiing was primarily developed as sit skiing and was designed for individuals with mobility or lower-body impairments. The skier uses a mono-ski or bi-ski with a molded seat attached to the skier. The seat itself has undergone transformation during the past three decades from a one-size-fits-all seat to seats designed for different body types. Seats for high-performance skiers are molded to specifically fit the user. Technological advances have allowed the seats to evolve for more comfort and higher performance. Other technological innovations that have improved the user experience in sit skis are improvements in shock absorbers, suspension systems, and tethering systems that provide novice skiers or individuals with limited dexterity with the ability to ski with the assistance of another skier.

A key provider of sit skis for Challenge Aspen is Enabling Technologies, which is based in Denver, Colorado. The organization is focused primarily on alpine ski equipment and is an industry leader in innovation for adaptive skiing. Enabling Technologies was founded by T.G. "Woody" Witte, who focused on creating more advanced adaptive ski equipment for Ken LaCome, who was left paralyzed from a motorcycle accident in the early 1980s. Witte developed a high-performance mono-ski that gained national attention as LaCome was competing on the national circuit. Witte continued to use lightweight materials, advances in suspension systems, and weight transfer techniques to develop high-performance skis for both competitive and recreational use.

Tetradapt, based in Salt City, Utah, was founded at the University of Utah and is focused on using assistive technologies to overcome barriers for individuals with complex disabilities. Challenge Aspen also uses its equipment, including its flagship innovation, the TetraSki. TetraSki uses technology that allows users to independently control turning and speed with a joystick or sip-and-puff breath control. TetraSki is designed for individuals with limited strength and dexterity due to conditions such as spinal cord injuries, multiple sclerosis, or muscular dystrophy. Challenge Aspen creates life-changing opportunities, and technological innovations have enabled them to overcome barriers to provide those snowsport opportunities.

Summary

Adaptive sport and recreation provide opportunities for people with disabilities to compete in elite-level sports and participate in community recreational activities. Technology contributes to increased athletic performance, accessibility, and inclusion for athletes and participants at all levels. For elite athletes, technological advancements in sports-specific equipment, prosthetics, racing wheelchairs, and wearable performance trackers have enhanced performance. Additionally, immersive digital experiences, AI-driven training tools, and smart communication systems help athletes to be included and engage with fans in more interactive ways.

On the recreational side, technology has improved comfort, safety, and accessibility. Ultimately, making sure that individuals with disabilities can participate in recreational activities. Despite these advancements, challenges remain in awareness, cost, access, and fairness in competition. Many individuals, both with and without disabilities, are unaware of adaptive sports opportunities, leading to lower participation rates and financial support. Furthermore, the high cost of adaptive equipment and the lack of structured programs, particularly for individuals with complex disabilities, continue to restrict access. There is still a need for continual education regarding the value of adaptive sports and recreation and providing opportunities for participation.

Key Resources

- **Ossur**: https://www.ossur.com/en-us/neuro-orthotics
- **Liquid Access**: https://liquidaccess.org/
- **AcceSurf**: https://www.accessurf.org/
- **International Paralympic Committee "Raw Emotions"**: https://www.paralympic.org/raw-emotions
- **Project Guideline**: https://research.google/blog/project-guideline-enabling-those-with-low-vision-to-run-independently/
- **Interview with adaptive mountain biker Annijke Wade**: https://youtu.be/0imwYHixfaI
- **Challenge Aspen**: https://challengeaspen.org/
- **TetraSki**: https://www.tetraski.us/

References

Bakatchina, S., Weissland, T., Brassart, F., Alberca, I., Vigie, O., Pradon, D., & Faupin, A. (2022). Influence of wheelchair type on kinematic parameters in wheelchair rugby. *Frontiers in Sports and Active Living*, *4*, 861592. https://doi.org/10.3389/fspor.2022.861592

Brady, S. (2023). "A small leap for disabled man": The athlete-led evolution of the sports wheelchair and adaptive sports. *Sport in History*, *43*(1), 103–127. https://doi.org/10.1080/17460263.2022.2049634

Braza, D. W., Iverson, M., Lee, K., Hennessy, C., & Nelson, D. (2018). Promoting physical activity by creating awareness of adaptive sports and recreation opportunities: An academic–community partnership perspective. *Progress in Community Health Partnerships*, *12*(2), 165–172. https://doi.org/10.1353/cpr.2018.0036

Bridgestone Americas Press Center. (2024). Bridgestone drives premium performance through tires and beyond at the Olympic and Paralympic Games Paris 2024. Retrieved from https://www.bridgestoneamericas.com/en/press-release-details.en.2024.bridgestone-paris-2024-olympics-support

Bringing People Together Around the Paralympics. (2021). Retrieved from https://about.fb.com/news/2021/08/bringing-people-together-around-tokyo-paralympics/

Brittain, L. (2014). *From Stoke Mandeville to Sochi: A History of the Summer and Winter Paralympic Games*. Common Ground Publishing.

Broudy, B. (2023). *Adapt and Overcome*. Retrieved from https://www.adventurecycling.org/blog/adapt-and-overcome/

Bruce, W. (2023). Breaking barriers: The rise of adaptive mountain biking. Retrieved from https://www.singletracks.com/community/breaking-barriers-the-rise-of-adaptive-mountain-biking/

Cooper, R. A., & De Luigi, A. J. (2014). Adaptive sports technology and biomechanics: Wheelchairs. *PM & R*, *6*(8), S31–S39. https://doi.org/10.1016/j.pmrj.2014.05.020
de Groot, S., Gervais, P., Coppoolse, J. M., Natho, K., Bhambhani, Y., Steadward, R., & Wheeler, G. D. (2003). Evaluation of a new basketball wheelchair design. *Technology and Disability*, *15*(1), 7–18. https://doi.org/10.3233/TAD-2003-15103
De Luigi, A. J. (2024). The effects on sports performance of technologic advances in sports prostheses and wheelchairs. *PM & R*, *16*(4), 409–417. https://doi.org/10.1002/pmrj.13153
DeVillez, J., Sabato, J., & Fiedler, G. (2022). Could prosthesis use provide a competitive advantage in darts? *Prosthesis (Basel, Switzerland)*, *4*(2), 244–252. https://doi.org/10.3390/prosthesis4020024
Duvall, J., Satpute, S., Cooper, R., & Cooper, R. A. (2021). A review of adaptive sport opportunities for power wheelchair users. *Disability and Rehabilitation: Assistive Technology*, *16*(4), 407–413. https://doi.org/10.1080/17483107.2020.1767220
Giwnewer, U., Rubin, G., Friedman, A., & Rozen, N. (2020). User assessment of a novel suspension for a wheelchair—A prospective, randomized, double blind trial. *Geriatric Orthopaedic Surgery & Rehabilitation*, *11*, 2151459320983268. https://doi.org/10.1177/2151459320983268
Gordon, A. H., & De Luigi, A. J. (2020). Adaptive cycling. *Current Sports Medicine Reports*, *19*(7), 266–271. https://doi.org/10.1249/JSR.0000000000000728
Guttmann. L. (1952). On the way to an international sports movement for the paralysed. *The Cord*, *5*(3) 7–23.
Haegele, J. A., & McKay, C. (2025) Theoretical and social considerations of adaptive sport and recreation. In R. Hardin & J. R. Pate (Eds.), *Introduction to Adaptive Sport and Recreation* (pp. 1–12). Human Kinetics.
Hardin, R., & Pate, J. R. (Eds.). (2025). *Introduction to Adaptive Sport and Recreation*. Human Kinetics.
Hobara, H. (2014). Running-specific prostheses: The history, mechanics, and controversy. *Journal of the Society of Biomechanisms*, *38*(2), 105–110. https://doi.org/10.3951/sobim.38.105
Iverson, M., Ng, A. V., Yan, A. F., Zvara, K., Bonk, M., Falk-Palec, D., Sylla, C., Strickland, M., Preston, R., & Braza, D. W. (2021). Navigator role for promoting adaptive sports and recreation participation in individuals with disabilities. *American Journal of Physical Medicine & Rehabilitation*, *100*(6), 592–598. https://doi.org/10.1097/PHM.0000000000001533
Jamieson, A. R., & Wijesundara, H. D. (2025). A review of adaptive equipment and technology for exercise and sports activities for people with disabilities. *Disability and Rehabilitation: Assistive Technology*, *20*(1), 33–45. https://doi.org/10.1080/17483107.2024.2372323
Kazunas, A. (2024). Beyond the ride: Annijke Wade. Retrieved from https://www.bikemag.com/news/beyond-the-ride-annijke-wade
Khurana, R., Wang, A., & Carrington, P. (2021). Beyond adaptive sports: challenges & opportunities to improve accessibility and analytics. In *ASSETS 2021 - 23rd International ACM SIGACCESS Conference on Computers and Accessibility*. https://doi.org/10.1145/3441852.3471223
Kumar, T. (2023). Annijke Wade feels confidence in uncertainty. Retrieved from https://www.outsideonline.com/outdoor-adventure/biking/daily-rally-podcast-annijke-wade/
LA28 Paralympic Sports Starting Lineup. (n.d.). Retrieved from https://la28.org/en/newsroom/LA28-Paralympic-Games-Initial-Sport-Program-to-Feature-22-Sports.html

Lee, K. K., & Uihlein, M. J. (2019). Adaptive sports in the rehabilitation of the disabled veterans. *Physical Medicine and Rehabilitation Clinics of North America, 30*(1), 289–299. https://doi.org/10.1016/j.pmr.2018.08.001

Legg, D., & Darcy, S. (2025). Innovation in disability, sport and recreation. In R. Hardin & J. R. Pate (Eds.), *Introduction to Adaptive Sport and Recreation* (pp. 33–41). Human Kinetics.

Leicester, R. (2024). Paris Olympics embrace accessibility technology for visually impaired fans. *TechXplore*. Retrieved from https://techxplore.com/news/2024-08-paris-olympics-embrace-accessibility-technology.html

Mason, B. S., Porcellato, L., Van Der Woude, L. H. V., & Goosey-Tolfrey, V. L. (2010). A qualitative examination of wheelchair configuration for optimal mobility performance in wheelchair sports: A pilot study. *Journal of Rehabilitation Medicine, 42*(2), 141–149. https://doi.org/10.2340/16501977-0490

Matsuwaka, S. T., & Latzka, E. W. (2019). Summer adaptive sports technology, equipment, and injuries. *Sports Medicine and Arthroscopy Review, 27*(2), 48–55. https://doi.org/10.1097/JSA.0000000000000231

McCarty, K., Bixby, L., & Kennedy, W. (2025). Barriers and opportunities for access and inclusion. In R. Hardin, & J. R. Pate (Eds.), *Introduction to Adaptive Sport and Recreation* (pp. 15–32). Human Kinetics.

Moorman, A. (2025). Legal aspects of disability sport & recreation. In R. Hardin, & J. R. Pate (Eds.), *Introduction to Adaptive Sport and Recreation* (pp. 63–88). Human Kinetics.

Nagamatsu, A. (2022). The importance of patents for sports prosthetics manufacturers. *Journal of Paralympic Research Group, 18*, 97–109. https://doi.org/10.32229/parasapo.18.0_97

Paoli, A. G., & Robles, J. (2025). Adaptive youth sports. In R. Hardin, & J. R. Pate (Eds.), *Introduction to Adaptive Sport and Recreation* (pp. 177–186). Human Kinetics.

Pate, J. R., Hardin, R., Shapiro, D., & McKay, C. (2023). Inclusion, infusion, or confusion: Exploring how faculty address adaptive sport and recreation in the sport management classroom. *The Journal of Hospitality, Leisure, Sport & Tourism Education, 33*, 100458. https://doi.org/10.1016/j.jhlste.2023.100458

Rayes, R., Ball, C., Lee, K., & White, C. (2022). Adaptive sports in spinal cord injury: A systematic review. *Current Physical Medicine and Rehabilitation Reports, 10*(3), 145–153. https://doi.org/10.1007/s40141-022-00358-3

Samsung Spain Launches Smart Cap to Help Blind Swimmers Compete. (2016). Retrieved from https://www.sportsbusinessjournal.com/Daily/Issues/2016/05/19/Technology/samsungs-blind-cap-vibrating-cap-lets-paralympic-swimmers-know-when-to-flip-turn/

Stran, M., & Hardin, B. (2025). University of Alabama adapted athletics. In R. Hardin, & J. R. Pate (Eds.), *Introduction to Adaptive Sport and Recreation* (pp. 205–221). Human Kinetics.

Toyota Mobility Foundation. (2024). TMF supports the International Paralympic Committee's sport development initiatives. Retrieved from https://pressroom.toyota.com/tmf-supports-the-international-paralympic-committees-sport-development-initiatives/

U.S. Olympic & Paralympic Committee. (n.d.). About para sport. *Para College Sport*. Retrieved from https://www.usopc.org/para-college-sport/about-para-sport

United States Census Bureau. (2024). Anniversary of Americans with Disabilities Act: July 26, 2024. Retrieved from https://www.census.gov/newsroom/facts-for-features/2024/disabilities-act.html

Usma-Alvarez, C. C., Fuss, F. K., & Subic, A. (2011). Effects of rugby wheelchair design on output velocity and acceleration. *Procedia Engineering*, *13*, 315–321. https://doi.org/10.1016/j.proeng.2011.05.091

Van Phillips Flex-Foot Prosthetics. (n.d.) Retrieved from https://lemelson.mit.edu/resources/van-phillips

World Health Organization. (2024a). Assistive technology. Retrieved from https://www.who.int/news-room/fact-sheets/detail/assistive-technology

World Health Organization. (2024b). WHO and International Paralympic Committee team up to highlight power of assistive technology at Paris 2024 Paralympics. Retrieved from https://www.who.int/news/item/28-08-2024-who-and-international-paralympic-committee-team-up-to-highlight-power-of-assistive-technology-at-paris-2024-paralympics

World Health Organization. (n.d.). Disability. Retrieved from https://www.who.int/publications/i/item/9789240063600

PART IV

Technology and Built and Natural Environments

12

TECHNOLOGY, SPORT VENUES, AND EVENT MANAGEMENT

Angela I. Hayslett and Katharine M. Nohr

Introduction

Technology has become an integral part of the sports industry, revolutionizing not only on-the-field activities like officiating but also off-the-field operations such as stadium management and fan engagement (Jacobs, 2024). As demonstrated in this chapter, cutting-edge stadium designs outfitted with interconnected Internet of Things (IoT) systems, immersive digital fan experiences driven by artificial intelligence (AI), extended reality (XR), and blockchain are transforming how events are organized, delivered, and consumed. These advancements enhance the in-person and virtual spectator experience and improve back-end logistics such as venue security, data analytics, and sustainability efforts (Chen, 2024; Chicas, 2024; O'Sullivan, 2024; Shamsi, 2024; Walch, 2024).

Technological innovation continues to reshape the landscape of sport and recreation, particularly within venues and event management (Chicas, 2024). From real-time data monitoring and crowd flow optimization (Chicas, 2024) to immersive fan engagement (Chicas, 2024; Kittel et al., 2024) and secure ticketing systems enabled by blockchain (Agbozo & Hayawi, 2024; O'Sullivan, 2024), technology is transforming how sport experiences are designed, delivered, and managed. Sport organizations are increasingly leveraging tools such as connected devices (Chen, 2024), AI (Walch, 2024), blockchain platforms (O'Sullivan, 2024), and XR (Shamsi, 2024) to enhance operations and elevate experiences for fans and event organizers. These technologies help improve fan interaction, venue sustainability (PwC, 2024), security, and the accuracy of officiating (Walch, 2024). This chapter explores these applications and their broader implications, offering insights into how sport venue and event managers can adopt emerging technologies to optimize event outcomes and enrich the participant experience.

DOI: 10.4324/9781003588405-16

Background

For centuries, people have gathered in stadiums and arenas to witness athletic competitions, and the evolution of these spaces has reflected the changing needs of societies around the world (Kyle, 2014). In ancient Rome, the Colosseum was constructed using innovative engineering methods to seat large crowds, and in ancient Greece, early Olympic stadiums employed comparable architectural principles to accommodate large-scale athletic events (Evans, 2025). Expansion of the railroad in the 1800s gave sportsmen and spectators greater access to travel to sporting events (Betts, 1953). Once illuminated by gas lighting, smoke-filled venues were replaced with a more inviting and healthier environment upon the introduction of electric lighting (Betts, 1953). In the early 2000s, smartphones, social media, high-definition television, and high-tech graphics changed where and how fans could follow their favorite teams (Saunders, 2016). Venue managers competing with the at-home experience began including high-definition televisions and wireless internet inside stadiums (Saunders, 2016).

Frevel et al. (2022) describe how the invention of cameras has furthered technological developments for use to aid referees and to entertain fans with the use of instant replay (cf. Vannatta, 2011), the Hawk-Eye in cricket (Collins & Evans, 2011), and the Goal-line technology in football (Winand & Fergusson, 2016). Now, video-assisted refereeing is helping to achieve more accurate refereeing results (Spitz et al., 2020). Today's spectators expect more efficient operations with digital ticketing, interactive seating charts, and mobile ordering that have streamlined stadium operations and enhanced the user experience (Orderific, 2023), implying the industry's commitment to growing and modernizing sports venues and event management. Real-time data analysis powered by AI helps event organizers streamline operations, from ticketing procedures (e.g., dynamic pricing based on demand) to game scheduling (Sports Venue Technology, 2025c). Data analytics tools further recognize patterns that enable proactive problem-solving, allowing event managers to adapt rapidly to changing circumstances (Vaiyyali, 2024).

These innovations are not solely about efficiency; they also deepen the connection between fans and events by creating personalized, interactive experiences, such as custom in-app features, targeted promotions, and real-time engagement opportunities (Sports Venue Technology, 2025c). The examples from this section show the influential role technological advances have played in shaping access, enjoyment, reporting, and analysis of sport. As this historical overview has demonstrated, ongoing technological progress will continue to shape venue design and event management practices.

Topical Exploration and Key Concepts

Technical innovations such as AI, blockchain, and XR have become a pivotal force in sport venues and event management, fundamentally reshaping how events are planned, monitored, and experienced (Chicas, 2024; Walch, 2024). The following sections identify some of the ways in which technology is impacting venue and event management, fan experience, and refereeing (Magaz et al., 2024).

Venue and Event Management

AI for Venue and Event Management

AI supports venue operations and event management, like the NBA's use of AI models to create a game schedule that balances the competing demands to increase viewership while minimizing player travel fatigue (Vina, 2025). Technology propels sustainability in sports events, such as in venues that rely on AI-driven energy management systems to adjust lighting, heating, and cooling in response to real-time crowd or climate conditions (Popke, 2023a). The National Football League (NFL) franchise, the New England Patriots, use AI to manage team logistics and automate utilities based on usage (Bradley, 2025). The same analytical engines extend to security, where real-time crowd-monitoring platforms fuse facial-recognition data with motion-sensor alerts to flag anomalies before they escalate (Sports Venue Technology, 2025c).

Behind the scenes, AI-driven predictive-maintenance dashboards continuously scan structural, mechanical, and electrical systems, enabling stadium operators to intervene long before small faults become costly shutdowns (Sports Venue Technology, 2025c). Energy management too has become data-centric: smart building controls learn occupancy patterns to fine-tune lighting, heating, ventilation, and air conditioning (HVAC) units, and water use, thereby lowering both costs and the facility's carbon footprint (Sports Venue Technology, 2025c).

On the revenue side, dynamic-pricing algorithms adjust ticket values in real time, harmonizing supply with fluctuating demand and reshaping seating layouts to boost yield per square foot (Sports Venue Technology, 2025c). Concessionaires gain similar foresight as inventory planning models forecast food, beverage, and merchandise needs down to the stock keeping unit (SKU), reducing waste while preserving stock availability (Sports Venue Technology, 2025c). Finally, frictionless mobile ordering, guided by AI prediction of queue lengths and purchase intent, cuts wait times and elevates satisfaction, and post-purchase analytics help marketers deliver hyper-targeted in-stadium offers that increase customer spending (Sports Venue Technology, 2025c). One example of the use of AI for mobile concessions is found at Gillette Stadium, home to the New England Patriots (Bradley, 2025). When AI is combined with enhanced features like digital ticketing and other contactless payment systems, it can reduce waste, reduce energy consumption, and lower costs (Sports First, 2025).

Safety measures in large venues benefit from the same AI that revolutionizes performance analytics and event management. Generative AI algorithms can sift through video feeds, detect crowd anomalies, and guide security staff to potential threats (Popke, 2023a, 2023b). Facial recognition technology for staff, players, and credentialed media personnel reduces the chance of unauthorized access (Popke, 2023b). AI can also sift through vendor data and vet personnel as well as analyze policies and procedures (Popke, 2023a). Even weapon-detection sensors have grown more sophisticated, alerting security to concealed threats and reducing the need for invasive searches.

Blockchain for Security and Transparency to Safeguard Event Transactions

Among these transformative tools, blockchain plays a prominent role in verifying transactions and safeguarding sensitive information. Defined as "an open distributed ledger," blockchain helps venue managers track event logistics, from equipment movement to digital transactions (Agbozo & Hayawi, 2024, p. 158). It is nearly impossible to alter or tamper with the data, and the transparency helps to maintain integrity and trust among stakeholders (Agbozo & Hayawi, 2024).

Blockchain technology can also help secure copyrights with anti-counterfeiting codes attached to unique timestamps linked to sporting events (Agbozo & Hayawi, 2024). This feature protects intellectual property and ensures that content creators receive proper recognition and compensation for their work. Additionally, cryptocurrencies, which use blockchain technology, can enhance safety in sports betting by providing secure and transparent transactions (Agbozo & Hayawi, 2024). When used alongside IoT, machine learning, and AI, blockchain technology further automates ticketing and financial processes, streamlining back-end tasks that once required manual oversight (Agbozo & Hayawi, 2024).

Risk Management and Cybersecurity to Reduce Event Risk

In tandem with these ledger-based tools, risk management technologies have become indispensable for large-scale sporting events (Hall & Allen, 2024). Mobile applications and social media can deploy critical messages to spectators and emergency professionals, ensuring that everyone is informed and prepared for adverse weather conditions (Hall & Allen, 2024).

Drone detection systems alert organizers to unauthorized drones that might be weaponized or used for illicit surveillance (Hall & Allen, 2024). Gunshot detection software uses sound acoustics to distinguish between actual gunshots and similar sounds, and explosive particle detectors can help identify explosive devices (Hall & Allen, 2024). Biometric systems like facial recognition can enhance security (Sports Venue Technology, 2025b). Other security measures include advanced scanners, retractable bollards, and biometric entry systems designed to thwart terrorist threats or active shooters and ensure access to properly credentialed individuals (Hall & Allen, 2024). These technologies enhance the security of events by mitigating threats, providing real-time alerts, and enabling quick responses to potential threats.

Cybersecurity remains a top priority, given the volume of digital transactions and data-sharing. Systems such as Security Information and Event Management (SIEM) continuously analyze network logs to detect unusual activity before it escalates into a breach (Sharif, 2025). By integrating risk management and cybersecurity tools, venues significantly enhance their resilience against a wide range of potential disruptions.

Digital Twins for Venue Management

Event organizers are increasingly using digital twins, which are virtual replicas of physical assets, environments, and processes that enable real-time monitoring, simulation, and optimization of resources (Glebova et al., 2023). By modeling real-world layouts and conditions, digital twins are transforming how events are designed, managed, and executed (The Imagination Collaborative, 2024). With a virtual version of the event space, with layouts, audience movement, lighting, sound, and even weather impacts, event planners can foresee issues, test scenarios, and make informed decisions long before the actual event takes place (Glebova et al., 2023). Digital twins can simulate various strategies for waste reduction or alternative energy usage, guiding venue operators toward eco-friendly choices (Glebova et al., 2023).

Moreover, digital twins support day-of-event adaptability by incorporating real-time IoT data, allowing for adjustments in real time (The Imagination Collaborative, 2024). Digital twins can optimize utilities and energy consumption, contributing to event sustainability. One of the key applications of AI is the creation of digital twins, which help in simulation planning and training for real-world scenarios (Popke, 2023a). By simulating various scenarios, event managers can anticipate potential risks and develop strategies to mitigate them (Popke, 2023a). Organizers can also use digital twins to measure an event's success and improve future events by analyzing fan engagement and eliciting feedback and surveys (Glebova et al., 2023). A digital twin environment creates another realm for advertising and sponsorship, strengthening brand identity and relationships while promoting sustainable practices (Glebova et al., 2023). By combining these measures, sport organizations can lower their carbon footprint while preserving fan comfort and operational efficiency (Glebova et al., 2023).

Drone, Robotic, and Hoverbike Use for Risk Management

Aerial drone footage can help event managers assess environmental hazards, traffic congestion, or crowd surges before they become critical problems (Nohr & Hayslett, 2025). They can assist with route planning by creating 3D maps or virtual reality models, enhancing preparation and safety (Seidaliyeva & Smailov, 2025). Drones can be used to scout hazards, giving time for early intervention by organizers to enhance event safety (Seidaliyeva & Smailov, 2025).

By surveying the venue's perimeter and interior, drones supply real-time images that human spotters simply cannot match. Sporting events in remote or large-scale settings especially benefit from this aerial surveillance, as it can locate issues from extreme weather changes to unauthorized gatherings with greater accuracy (Nohr & Hayslett, 2025). By providing a bird's-eye view of the event, drones can help event managers make informed decisions about crowd control and emergency response (Nohr & Hayslett, 2025). Furthermore, drones can assist with providing and evaluating a prompt medical response to injured persons (Sport Drones, 2025).

Robotics and hoverbikes can be used in sport venues that are complex and spread out (Nohr & Hayslett, 2025). They can be outfitted with cameras, sensors, and even transport people to more readily navigate to an emergency at an event (Nohr & Hayslett, 2025). These futuristic-looking devices can be quickly and widely deployed (Nohr & Hayslett, 2025). These technologies enhance the ability to ensure the safety of participants and spectators, making events more secure and enjoyable (Nohr & Hayslett, 2025).

Wearable Sensors and Emergency Communication

Wearable sensors can bolster event-wide safety. Devices that detect abnormal head impacts (Tyson et al., 2018) or measure crowd density can dispatch automated alerts to medical or security teams (Nohr & Hayslett, 2025). In crisis situations such as severe weather or a fire, these wearables can also coordinate evacuation routes by sending directions to staff and attendees in real time (Nohr & Hayslett, 2025). Wearables can also be used for communication in a crisis to relay timely information to key event personnel in the event of a man-made or environmental danger (Nohr & Hayslett, 2025). Strategic information can also be deployed to athletes and fans on their wearables (Nohr & Hayslett, 2025).

Three-Dimensional Laser Scanning to Maintain a Safe Facility

Three-dimensional (3D) laser scanning ensures that stadiums and arenas remain structurally sound (Gura et al., 2021a). By generating high-resolution, three-dimensional images, this technology detects potential surface deformities, such as uneven running tracks, before they pose hazards to athletes or spectators (Gura et al., 2021b). These scans can integrate with facility management software to schedule maintenance, monitor seasonal wear, and confirm that any renovations maintain optimal safety standards (Gura et al., 2021a).

Unlike spot surveys that sample only a handful of coordinates, a single scan records the entire venue, giving operators a comprehensive baseline against which the smallest deformation or corrosion can be detected over time (Gura et al., 2021b). Because every scan is georeferenced, operators can repeat the process on a quarterly or annual cycle and auto-register new point clouds against the original model (Gura et al., 2021a). Any divergence appears as a color-graded heat map that prioritizes repairs (Gura et al., 2021a). World Athletics now accepts LiDAR flatness reports when certifying track surfaces for elite competition, and university facilities have used the same technique to confirm that roughness stays below the 10-mm threshold required for accreditation (Gura et al., 2021a).

Major event organizers rely on laser scanning to compress construction timelines while guaranteeing as-built quality. During the build-out of Beijing's National Speed-Skating Oval ("Ice Ribbon"), engineers scanned the ice sheet and its sub-floor piping network at every construction milestone, saving an estimated 4% in costs (Zhang et al., 2022). In the United States, SoFi Stadium was scanned in just two on-site days ahead of Super Bowl LVI (GPRS, 2025a).

Laser scanning is also used for routine surface-quality and accessibility audits. Sportworks Design deploys LiDAR on synthetic-turf, tennis-court, and track replacements to generate elevation heat maps that identify high or low spots exceeding requirements (Hammond, 2025). Many venue managers now scan ramps and concourses to confirm compliance with the Americans with Disabilities Act (ADA) slopes to within ±2 mm, which is difficult using manual tape and level readings (GPRS, 2025c). When Crypto.com Arena added 3,000 linear ft of LED ribbon boards, contractors used the arena's scan in their structural analysis software to design anchor points without guesswork, eliminating change orders and ensuring that the heavy display did not exceed live-load limits (GPRS, 2025b).

Fan Experience

Digital Systems to Streamline the Fan Experience

Seamless entry and ticketing solutions encompass mobile ticket delivery, biometric verification, and automated gate systems aimed at minimizing bottlenecks and improving the stadium experience (Popke, 2023b). Electronic ticket scanning has become the norm, cutting down on fraud and simplifying crowd control (Popke, 2023b). However, by integrating facial authentication with electronic ticket scanning, venues can match a fan's face to their digital ticket in seconds, which not only reduces fraud but also creates a personalized welcome message or promotional offers on a user's mobile app (Popke, 2023b). The Intuit Dome arena, which is home to the NBA's Los Angeles Clippers, uses facial authentication technology for entry and concession payments, which they call "Game Face ID" (Kapustka, 2024).

Digital seating allows fans to upgrade seats while preventing overcrowding (Sports First, 2025). Moreover, real-time seat allocation can optimize venue capacity by dynamically adjusting seat availability based on ongoing attendance data (Sports First, 2025). This technology improves the entry process and ensures that only authorized individuals can access restricted spaces, enhancing security and convenience (Sports First, 2025). As these solutions continue to evolve, privacy safeguards, such as data encryption, limited retention policies, and user consent protocols, are increasingly essential to maintain fan trust (Popke, 2023a).

Contactless payment systems streamline transactions, reduce wait times, and enhance the overall fan experience (Sports First, 2025). These systems are particularly valuable in high-traffic areas such as concessions and merchandise stands (Sports First, 2025). Beyond convenience, contactless payments also support advanced loyalty programs, where each purchase can automatically accumulate points or offer targeted discounts in real time (Sports First, 2025). In addition, they minimize cash handling, which can reduce theft and enhance health and safety measures in high-volume environments (Sports First, 2025).

Blockchain Technology to Secure Ticketing and Financial Data

Blockchain technology further eliminates counterfeit tickets by logging each step of a ticket's lifecycle on a decentralized ledger (Agbozo & Hayawi, 2024). Additionally, blockchain can manage digital trading cards (Chen et al., 2022) and virtual tokens, providing fans with access to exclusive offers (Vidal-Tomas, 2023) and governance decisions (Stegmann et al., 2023). Sports teams can use blockchain for their loyalty programs, fostering a deeper connection between fans and the sports organizations they support (Choudhary, 2024). Through tokenization, fans can collect unique merchandise (Vidal-Tomas, 2023), participate in voting on team-related decisions, or even have a say in minor operational aspects, thus deepening their sense of involvement and loyalty (Stegmann et al., 2023).

Blockchain technology can aid in fantasy leagues to trade assets or in sports betting transparency and fairness by validating transactions (Choudhary, 2024). It can also be used to authenticate memorabilia with digital certificates (Choudhary, 2024). Once authenticated, these certificates can be instantly verified by third parties, enhancing consumer confidence in sports collectibles (Choudhary, 2024). Teams can also offer a decentralized streaming service through blockchain, where fans can pay to access content (Choudhary, 2024). Such services can include tiered subscription models, allowing fans to pay for premium features like behind-the-scenes access, exclusive interviews, or archived classic matches (Choudhary, 2024).

Mobile Integration and Smart Stadiums to Enhance the Fan Experience

Smart stadiums can tip the scale for in-person entertainment by offering advanced technological amenities (Sports Venue Technology, 2025a). Inside the venue, high-speed 5G networks improve connectivity for fans using mobile apps to order concessions, access real-time game stats, or find the shortest restroom lines (Sports Venue Technology, 2025a). These smart stadium initiatives often include large 4K screens for live replays (Patel, 2024), while Wi-Fi 6 deployments cater to the higher data usage in big crowds (Intel, n.d.).

Stadium operators can also use location-based services to send personalized push notifications about seat upgrades, loyalty rewards, or interactive fan challenges (Sports Venue Technology, 2025a). Augmented reality (AR) content may also appear on a fan's mobile device, overlaying player avatars or 3D graphics onto the real-world game (Patel, 2024). By providing fans with access to advanced amenities, smart stadiums encourage attendance and create memorable experiences. In addition, smart-stadium dashboards can aggregate data analytics from foot traffic, concession sales, and fan engagement metrics to optimize staffing and concession inventory in real time (Patel, 2024).

XR and Digital Twins to Entertain Fans

The use of XR and digital twins not only benefits sport venue and event managers but also broadens possibilities to enhance the fan experience (Catalin et al., 2023; McCaskill, 2024). By merging physical events with digital twins, spectators can engage with a replica of the venue to navigate busy corridors, minimize concession wait times, and even choose from multiple viewing angles in near real time (Glebova et al., 2023). Digital twins can also incorporate crowd density and traffic flow data, allowing fans to plan efficient routes or locate the quickest concession stand for faster service (Glebova et al., 2023).

Those unable to attend in person can immerse themselves in hyper-realistic VR environments (Catalin et al., 2023), complete with interactive fan zones and athlete storytelling experiences (Glebova et al., 2023). XR "has the potential to redefine fan engagement by allowing fans to see statistical overlays and data analysis, have virtual athlete meetups, and even shop and try on merchandise in real time" (Patel, 2024, AR/VR, para. 1). For sponsors and advertisers, XR also opens new revenue streams by offering immersive brand activations that resonate with both on-site and remote fans (Patel, 2024).

Refereeing

XR Referee Training Simulations

Refereeing has undergone its own transformation with technology (Kittel et al., 2024). XR platforms, including VR-based training simulations, aim to sharpen officials' perceptual-cognitive skills (Kittel et al., 2024). While VR cannot fully replicate on-the-move officiating, it does recreate many high-pressure scenarios, including penalties, fouls, and offside rulings that demand repositioning for rapid judgment (Kittel et al., 2024).

A systematic review of 64 studies reported significant gains in anticipation and decision time among officials who trained with XR compared with those using two-dimensional video (Kittel et al., 2024). First-person 360° VR video delivered through head-mounted displays preserves the continuous optic-flow cues officials rely on in live play, thereby giving perceptual-cognitive drills much higher ecological validity than standard 2D broadcast footage (Boyer et al., 2023; Kittel et al., 2024). Experimental work with sub-elite football referees shows that decisions made inside a 360° VR environment predict on-field accuracy more closely than those made with flat-screen video and shorten clip-to-decision latency by about 15% (Kittel et al., 2024). Early tournament data suggest that these overlays reduce positioning errors by roughly 30% (MacInnes, 2024).

Video Assistant Referees (VAR)

Video Assistant Referee (VAR) systems have become prominent in soccer and other sports, reducing controversial calls and increasing the accuracy of game-altering decisions (Spitz et al., 2020). By giving referees immediate access to replays from multiple angles, VAR limits human error and fosters a greater sense of fairness (Spitz et al., 2020). VAR stations an off-field crew who can review potential match-changing incidents in real time through multiangle high-frame-rate feeds (Spitz et al., 2020). An analysis of 2,195 elite matches across 13 leagues showed that VAR interventions raised final-call accuracy from 92% to 98%, while keeping median check duration to 22 seconds (Spitz et al., 2020).

Semi-automated off-side technology (SAOT), which debuted at the 2022 FIFA World Cup, combines 12 roof-mounted tracking cameras with a sensor-equipped ball to map 29 data points per player and generate an off-side line automatically (FIFA, 2023). Observers at UEFA Euro 2024 report that SAOT trims about 30 seconds from the typical 70-second VAR off-side review (MacInnes, 2024). Following successful testing, the English Premier League planned to use SAOT in 2025, citing gains in "speed, efficiency and consistency" (Premier League, 2025, para. 3).

IOT: Wearable Devices for Improved Referee Outcomes

Referees also benefit from the IoT, such as wearable technology (Pan et al., 2022). Smart patches and wristbands can track officials' positioning, leading to more consistent officiating across matches (Pan et al., 2022). Wearables can further record how quickly a referee signals fouls or confirm the accuracy of hand signals, providing feedback to refine performance in subsequent matches (Pan et al., 2022). Training suites that pair surface electromyography (sEMG) with inertial measurement units now quantify the speed and clarity of referees' hand-signal mechanics, replacing coach-eyeballed assessments with objective scores (Pan et al., 2022). A hierarchical recognition algorithm based on these sensor streams has achieved gesture-classification accuracies above 95%, enabling automated feedback after each drill (Pan et al., 2022).

Smartwatches with specially designed apps such as REFSIX (2022) can help referees record key game incidents like cautions, goals, substitutions, as well as tracking game time. The watch can track referee fitness and positioning by tracking distance, game speed, and sprinting (REFSIX, 2022). Wearables with global positioning system (GPS) trackers help officials monitor their coverage of the field (Refr Sports, 2024). Devices that monitor heart rate variability and post-match recovery can help sport officials ensure they are physically and mentally fit for matches, reducing fatigue-based mistakes (Refr Sports, 2024).

The Future of Technology in Sport Venues and Event Management

According to Magaz et al. (2024), technology's role in sport venues and event management will continue to expand, with each innovation opening new avenues for efficiency, engagement, and security. This section suggests possible futures of the intersection of technology in sport venues and event management. To start, the rise of AI-driven predictive analytics, biometric security systems, advanced energy-management solutions, and immersive AR/VR experiences shows no sign of slowing. As these tools develop, they will likely become more user-friendly, cost-effective, and integrated across multiple aspects of sport. Practitioners can navigate this evolving landscape by taking the following steps. First, they should thoroughly assess their venue's needs and existing systems. Second, they should research and experiment with relevant technologies, ensuring that staff training is a cornerstone. Third, they should monitor each new system's performance, gathering feedback from athletes, officials, and fans, to enable ongoing refinement. Finally, practitioners should stay informed about emerging trends and attend industry events or webinars to ensure that decision-makers keep pace with the latest advancements.

By adopting a forward-thinking, research-driven approach, sport organizations can harmonize operational excellence with innovative and inclusive experiences for fans and participants alike. This holistic integration of technology from wearables and drones to blockchain and XR promises a safer, greener, and more engaging future for sporting events worldwide.

Trends and Current Issues

Sport venue innovation has entered a systems phase in which once separate digital tools, such as network infrastructure, point-of-sale terminals, security cameras, HVAC sensors, and broadcast control rooms, converge onto a single, cloud-orchestrated platform that can be reconfigured for each event (Sports Venue Technology, 2025c). This "venue operating system" approach means that a football stadium can morph overnight into a concert hall or esports arena, with lighting cues, concession menus, and ticket gates updated through software rather than hard wiring (Glebova et al., 2023). The payoff is operational agility, but the dependency on tightly coupled systems also raises the stakes for downtime and cyberattack (Sharif, 2025).

As the data streams multiply, governance and privacy have become board-level concerns because biometric entry, AI video analytics, and fan-engagement apps now collect sensitive personal information at the industrial scale (Magaz et al., 2024). Venues are adopting zero-trust architectures, real-time log analysis, and third-party penetration testing to satisfy regulators and insurers that data flows remain auditable and consent-based (Sharif, 2025). Parallel discussions around algorithmic bias are prompting ethics audits of facial recognition systems to ensure equal treatment of diverse fan populations (Marks, 2025).

Lenders now demand proof that billion-dollar stadium builds can withstand extreme heat, flooding, or wildfire smoke (Popke, 2023a). Digital twin simulations are now run months before groundbreaking to model solar panel output, water recycling loops, and evacuation routes under multiple climate scenarios, creating an evidence trail for green bond issuances and municipal approvals (Glebova et al., 2023). AI energy-management dashboards use HVAC and LED ribbon boards to decrease carbon intensity during off-peak occupancy inside arenas (Sports Venue Technology, 2025b).

Fans have shifted from passive ticket buyers to token-holding stakeholders who vote on kit designs or charity partners via blockchain platforms (O'Sullivan, 2024). Tokenized loyalty schemes also give clubs real-time insight into purchasing behavior while granting fans portable, resale-proof digital tickets and merchandise certificates (Choudhary, 2024).

The hyper-connected stadium is also a cyber-physical target, with hostile drones, deepfake evacuation messages, and ransomware campaigns now listed in venue risk registers alongside crowd crush and lightning strikes (Hall & Allen, 2024). Scenario planning, therefore, pairs drills with tabletop exercises that run simulated firewall breaches or rogue IoT accelerometer spikes so operators can rehearse coordinated responses across IT, facilities, and public safety teams (Sharif, 2025).

Finally, the fusion of technology and entertainment is reshaping the workforce (Marks, 2025). Broadcast engineers are retraining as real-time graphics programmers, and ushers wielding handheld scanners now troubleshoot edge compute nodes that feed predictive algorithms (Marks, 2025). Venue operators are now competing with tech startups for data-literate staff (Catalin et al., 2023). These examples suggest that the future of technology will continue to shape sports venues and event management.

Practice in Action

In the heart of Los Angeles, SoFi Stadium is a state-of-the-art facility that uses technology to increase operational efficiency and redefine the fan experience. The LA Rams, as one of its primary occupants, embody a spirit of innovation that leverages technology to engage fans and enhance performance. This section provides information from interviews with P. Maldonado, VP of IT Infrastructure & Operations at Hollywood Park and SoFi Stadium, and K. Kai-ling Frederick, Chief Marketing Officer, Los Angeles Rams.

SoFi Stadium Technology

SoFi Stadium's design and infrastructure are a masterclass in technological integration. While older stadiums often have separate networks for broadcast, security, HVAC, and concessions, SoFi Stadium merges these systems into one converged network and data center. This approach simplifies management and lets the stadium's IT team adapt swiftly to everything from NFL games to concerts and global events.

Infinity Screen and Real-Time Data

A centerpiece of this innovation is the Infinity Screen, a colossal 120-yard-wide, four-story-tall 4K display that fundamentally changes the way fans watch live sport. Beyond traditional replays, the Infinity Screen delivers real-time fantasy stats, out-of-town scores, and even a noise meter to ramp up crowd energy. Because it is fully integrated with back-end systems, the screen can zoom in on a game-changing foot placement or dynamically pull in data on a player's route running speed.

Reliability and Security

Events on the scale of Super Bowls, FIFA World Cup matches, or the Olympic Games demand bulletproof infrastructure. SoFi Stadium's redundancy is comprehensive: dual internet providers, duplicate firewalls, and hardened cybersecurity protocols help ensure that everything from lighting to point-of-sale systems remains operational, even if a hardware failure or cyber threat occurs. The stadium's team regularly runs large-scale tests, powering down sections of the network to guarantee instant failover and uninterrupted service.

The Los Angeles Rams and Innovation

The LA Rams, as a franchise of the NFL, are leveraging technology to cultivate fandom and enhance the game-day experience.

Data-Driven Marketing and Personalized Experiences

Behind the scenes, the Rams use data analytics to understand fan demographics and behaviors. This enables tailored outreach, like distributing youth-focused content on social media or highlighting fantasy football stats. The Rams also explore ways to integrate AI for operational efficiencies, from simple chatbots that answer fan FAQs online to creative tools that can build out marketing content in multiple formats. They remain cautious, however, about offloading player-driven storytelling to AI, mindful of inclusivity, authenticity, and player consent.

Augmented Reality and Interactive Apps

While the Infinity Screen draws eyes in the bowl, the Rams have expanded fan engagement to smartphones. SoFi Stadium was one of the first venues to incorporate Snap's in-venue camera kit for real-time AR interactions. By pointing a phone at the field, fans unlock immersive AR effects, touchdown celebrations, callouts for specific players, and even remote access for those watching from home. The same AR tools can instantly identify and spotlight celebrities in private suites, allowing for spontaneous crowd engagement without requiring them to move from their seats. These experiences cultivate a sense of community well beyond the stadium's physical walls.

Balancing Tradition with Modernization

From the iconic soundtracks that kick off each quarter to the newest frictionless concessions, the Rams blend longtime football rituals with cutting-edge fan experiences. The goal is not to replace beloved traditions, such as face painting or cheering in unison, but to elevate them so that fans of every generation feel the stadium's energy.

A Glimpse Ahead

As facial recognition and AI solutions become more widespread, SoFi Stadium is poised to explore them in ways that streamline security, speed up entry, and personalize fan interactions. Already, frictionless concessions and advanced point-of-sale systems are proving how foundational technology can revolutionize stadium operations. With the NFL's biggest events, the 2026 FIFA World Cup, and the 2028 Los Angeles Olympic Games on the horizon, SoFi Stadium's proactive, scalable IT model sets a benchmark for large venues worldwide.

Summary

Technological innovations, from IoT and AI to blockchain and XR, are fundamentally changing the sports venue and event landscape. These tools improve efficiency in sports venues and event management, enhance security, and create richer experiences for fans. As new developments continue to emerge, practitioners must remain adaptable and proactive, integrating the right solutions to meet organizational goals and audience expectations. By doing so, they can strengthen sustainability efforts, streamline operations, and ensure that sports events remain engaging, equitable, and safe for all involved.

Looking ahead, ongoing technological progress will continue to shape venue design and event management practices. As new ideas and digital tools emerge, the industry is poised to continue to transform the way sports events are conceived, organized, and enjoyed (The Imagination Collaborative, 2024).

Key Resources

- **International Association of Venue Managers (IAVM)**: https://iavm.org/
- **SportTechie**: https://www.sporttechie.com/
- **Green Sports Alliance**: https://greensportsalliance.org/
- **Stadium Tech Report**: https://stadiumtechreport.com
- **Sports Business Journal & Technology**: https://www.sportsbusinessjournal.com/Technology
- **Front Office Sports & Technology**: https://frontofficesports.com/category/technology

References

Agbozo, E., & Hayawi, W. M. (2024, March). A bibliometric overview of blockchain technology in sports. *Facta Universitatis Series Electronics and Energetics*, *37*(1), 157–168.

Betts, J. R. (1953, September). The technological revolution and the rise of sport, 1850–1900. *The Mississippi Valley Historical Review*, *40*(2), 231–256. https://doi.org/10.2307/1888926

Boyer, S., Rochat, N., & Rix-Lièvre, G. (2023, March 29). Uses of 360° video in referees' reflectivity training: Possibilities and limitations. *Frontiers in Psychology*, *14*. https://doi.org/10.3389/fpsyg.2023.1068396

Bradley, T. (2025, April 16). From stadiums to sidelines: How AI is reinventing the future of sports. *Forbes*. https://www.forbes.com/sites/tonybradley/2025/04/16/from-stadiums-to-sidelines-how-ai-is-reinventing-the-future-of-sports/

Catalin, P. M., Gabriel, P. M., Laurentiu, D. D., Daniel, P. L., & Eugen, C. D. (2023). Adoption of technological innovations by sports spectators. *Annals of the Constantin Brancusi University of Targu Jiu, Economy Series*, *2*, 151–160. https://www.utgjiu.ro/revista/ec/pdf/2023-02/15_Popescu_Popa%20_Diaconescu_P%C4%83s%C4%83rin_Costin.pdf

Chen, C.L., Fang, C.C., Zhou, M., Tsaur, W.J., Sun, H., Zhan, W., & Deng, Y.Y. (2022). A blockchain-based anti-counterfeit and traceable NBA digital trading card management system. *Symmetry*, *14*(9), 1827. https://doi.org/10.3390/sym14091827

Chen, H. (2024). Advancements in human motion capture technology for sports: A comprehensive review. *Sensors and Materials*, *36*(7), 2705–2726. https://doi.org/10.18494/SAM5084

Chicas, J. (2024, December 24). Smart stadiums: Redefining fan experience in sports. *Applaudo*. https://applaudo.com/es/insights/articles/smart-stadiums-redefining-fan-experiences-in-sports/

Choudhary, A. (2024, November 19). Blockchain in sports: Use cases, advantages and challenges. *A3Logics*. https://www.a3logics.com/blog/blockchain-in-sports/

Collins, H., & Evans, R. (2011). Sport-decision aids and the "CSI-effect": Why cricket uses Hawk-Eye well and tennis uses it badly. *Public Understanding of Science*, *21*(8), 904–921. https://doi.org/10.1177/0963662511407991

Evans, F. (2025, May 28). How the Colosseum was built – and why it was an architectural marvel. *A&E Television Networks*. https://www.history.com/articles/how-roman-colosseum-built

FIFA. (2023, July 17). Semi-automated offside technology. *Inside FIFA*. https://inside.fifa.com/innovation/womens-world-cup-2023/semi-automated-offside-technology

Frevel, N., Beiderbeck, D., & Schmidt, S. L. (2022). The impact of technology on sports – A prospective study. *Technological Forecasting & Social Change*, *182*. https://doi.org/10.1016/j.techfore.2022.121838

Glebova, E., Book, R., Su, Y., Peric, M., & Heller, J. (2023, December). Sports venue digital twin technology from a spectator virtual visiting perspective. *Frontiers in Sports and Active Living*, *5*. https://doi.org/10.3389/fspor.2023.1289140

Ground Penetrating Radar Systems (GPRS). (2025a). *3D Laser Scanning SoFi Stadium for 2022 Super Bowl Mixed Reality*. https://www.gp-radar.com/case-studies/3d-laser-scanning-sofi-stadium-for-2022-super-bowl-mixed-reality#:~:text=GPRS%20laser%20scanned%20SoFi%20Stadium,model%20for%20augmented%20reality%20use

Ground Penetrating Radar Systems (GPRS). (2025b). *GPRS Laser Scans Stadiums and Arenas: Precise as Built Data Delivered for Signage and LED Video Display Installation*. https://www.gp-radar.com/article/gprs-laser-scans-stadiums-and-arenas

Ground Penetrating Radar Systems (GPRS). (2025c). *Guide to 3D Laser Scanning for Concrete Construction*. https://www.gp-radar.com/article/guide-to-3d-laser-scanning-for-concrete-construction

Gura, D. A., Bespyatchuk, D. A., Samarin, S.V., Kiryunikova, N. M., & Lesovaya, E. D. (2021a, August). Technology of three-dimensional laser scanning as a tool to provide safety for sport facilities. *Nanotechnologies in Construction*, *13*(4), 259–263. https://doi.org/10.15828/2075-8545-2021-13-4-259-263

Gura, D. A., Bespyatchuk, D. A., Samarin, S. V., Kiryunikova, N. M., & Lesovaya, E. D. (2021b). Technology of three-dimensional laser scanning as a tool to provide safety for sport facilities. *Nanotechnologies in Construction*, *13*(45), 259–263. https://doi.org/10.15828/2075-8545-2021-13-4-259-263

Hall, S., & Allen, B. (2024, September 1). Current threats and emerging trends in sports safety and security. *Sports Destination Management*. https://www.sportsdestinations.com/management/safety-security/current-threats-and-emerging-trends-sports-safety-34659

Hammond, K. (2025). Take your synthetic turf field to the next level with 3D laser scanning. *Sportworks Design* https://www.sportworksdesign.com/take-your-synthetic-turf-field-to-the-next-level-with-3d-laser-scanning/

Intel. (n.d.) What is Wi-Fi 6? https://www.intel.com/content/www/us/en/gaming/resources/wifi-6.html

Jacobs, B. (2024, April 30). How will sports change in the future? What you should know. *American Public University*. https://www.apu.apus.edu/area-of-study/nursing-and-health-sciences/resources/how-will-sports-change-in-the-future/

Kapustka, P. (2024, August 16). Intuit Dome leaning on facial authentication for ticketing and concession transactions. *Stadium Tech Report*. https://stadiumtechreport.com/feature/intuit-dome-leaning-on-facial-authentication-for-ticketing-and-concession-transactions/

Kittel, A., Lindsay, R., Le Noury, P., & Wilkins, L. (2024). The use of extended reality technologies in sport perceptual-cognitive skill research: A systematic scoping review. *Sports Medicine*, *10*(128), 1–20. https://doi.org/10.1186/s40798-024-00794-6

Kyle, D. G. (2014). *Sport and Spectacle in the Ancient World*. John Wiley & Sons, Incorporated.

MacInnes, P. (2024, June 23). Euro 2024 innovations helping referees but some things technology cannot fix. *The Guardian*. https://www.theguardian.com/football/article/2024/jun/23/euro-2024-innovations-make-decisions-easier-but-human-element-remains

Magaz, A. M., Gallardo, L., Marin-Farrona, M., Sanchez-Sanchez, J., Lorenzo, A., Lopez-Fernandez, J., Duclos-Bastias, D., & Garcia-Unanue, J. (2024, March 7). Technological structure and configuration of the use of technology in European sports organizations. *Cultura, Ciencia y Deporte*, *19*(60), 147–160. https://doi.org/10.12800/ccd.v19i60.2145

Marks, R. (2025, April 4). Trends in entertainment venue management: A 2025 perspective. *Odgers Berndtson*. https://www.odgersberndtson.com/en-us/insights/trends-in-entertainment-venue-management-a-2025-perspective/

McCaskill, S. (2024, September 5). Olympic planning, immersive experiences and performance gains: How digital twins are recreating sport in the virtual world. *SportsPro*. https://www.sportspro.com/insights/features/digital-twins-sport-ai-analytics-oneplan-paris-2024-venues-metaverse-performance/

Nohr, K. M., & Hayslett, A. I. (2025). Application of modern technology for race course crisis communication. In B. E. Menaker, D. Sheptak, & J. J. Zhang (Eds.), *Crisis Management and Sports: Global Perspectives* (pp. 54–73). Routledge. https://doi.org/10.4324/9781003559214

O'Sullivan, J. (2024, July 15). The inside story of the football fan token community. *Forbes*. https://www.forbes.com/sites/josephosullivan/2024/07/15/the-inside-story-of-the-football-fan-token-community/

Orderific. (2023, December 4). History of sports facility management: Evolution unveiled. https://orderific.com/blog/history-of-sports-facility-management/

Pan, T. Y., Tsai, W. L., Chang, C. Y., Yeh, C. Y., & Hu, M. C. (2022, May). A hierarchical hand gesture recognition for sports referee training-based EMG and accelerometer sensors. *IEE Transactions on Cybernetics*, *52*(5), 3172–3183. https://doi.org/10.1109/TCYB.2020.3007173

Patel, R. (2024, September 18). Fan engagement in sports: Then vs. now. *FTI Delta*. https://www.ftidelta.com/insights/perspectives/fan-engagement-in-sports-then-vs-now

Popke, M. (2023a). Demystifying AI. *NCS4 Gameday Security Magazine*, 7–10. https://issuu.com/ncs4southernmiss/docs/gameday-fall-2023-digital-proof_1

Popke, M. (2023b). Stadium security 2.0. *NCS4 Gameday Security Magazine*, 9–12. https://issuu.com/ncs4southernmiss/docs/gameday-summer-2023-digital

Premier League. (2025, April 12). Semi-automated offside technology: What you need to know. https://www.premierleague.com/news/4256036

PwC. (2024). *Digital Stadiums: Transforming Sports and Fan Experiences with Technology in Saudi Arabia*. https://www.pwc.com/m1/en/publications/documents/2024/digital-stadiums-eng.pdf

Refr Sports. (2024, October 7). *How Wearable Technology Is Shaping Referee Performance: A Look at GPS, Accelerometers, and More*. https://www.refrsports.com/blog/how-wearable-technology-is-shaping-referee-performance-a-look-at-gps-accelerometers-and-more

REFSIX. (2022). *The App for Referees*. https://refsix.com/

Saunders, P. (2016, May 6). Decade in sports: High-tech era an instant hit. *The Denver Post*. https://www.denverpost.com/2009/12/26/decade-in-sports-high-tech-era-an-instant-hit/

Seidaliyeva, U., & Smailov, N. (2025). Leveraging drone technology for enhanced safety and route planning in rock climbing and extreme sports training. *Retos*, *63*, 598–609. https://doi.org/10.47197/retos.v63.110869

Shamsi, A. (2024, October 20). The growing fusion of technology and sports. *Reso*. https://www.resoinsights.com/insight/the-growing-fusion-of-technology-and-sports/

Sharif, A. (2025, March 5). Introduction to security information and event management (SIEM). *Crowdstrike*. https://www.crowdstrike.com/en-us/cybersecurity-101/next-gen-siem/security-information-and-event-management-siem/

Spitz, J., Wagemans, J., Memmert, D., Williams, A. M., & Helsen, W. F. (2020, August 14). Video assistant referees (VAR): The impact of technology on decision making in association football referees. *Journal of Sports Sciences*, *39*(2), 147–153. https://doi.org/10.1080/02640414.2020.1809163

Sport Drones. (2025, January 7). Why use drones for athlete performance metrics? https://sport-drones.com/technology-innovations-in-sport-drones/drones-for-measuring-athlete-performance-metrics/

Sports First. (2025). The evolving role of sports venue technology in venue management. https://www.sportsfirst.net/post/the-evolving-role-of-sports-venue-technology-in-venue-management

Sports Venue Technology. (2025a). 5G connectivity in sports venues: Revolutionizing the fan experience. https://www.sportsvenue-technology.com/articles/5g-connectivity-in-sports-venues-revolutionizing-the-fan-experience

Sports Venue Technology. (2025b). Future-proofing sports venues: Trends in technology adoption. https://www.sportsvenue-technology.com/articles/future-proofing-sports-venues-trends-in-technology-adoption

Sports Venue Technology. (2025c). Stadiums of tomorrow: How AI is enhancing sports venue operations. https://www.sportsvenue-technology.com/articles/stadiums-of-tomorrow-how-ai-is-enhancing-sports-venue-operations

Stegmann, P., Matyas, D., & Strobel, T. (2023, April 27). Hype or opportunity? Tokenization as engagement platform in sport marketing. *International Journal of Sports Marketing and Sponsorship*, *24*(4), 722–736. https://doi.org/10.1108/IJSMS-08-2022-0157

The Imagination Collaborative. (2024, September 18). *Operational Efficiency Using Digital Twins in the Event Industry: A New Era of Precision and Insight*. https://www.theimaginationcollaborative.com/post/operational-efficiency-using-digital-twins-in-events

Tyson, A. M., Duma, S. M., & Rowson, S. (2018). Laboratory evaluation of low-cost wearable sensors for measuring head impacts in sports. *Journal of Applied Biomechanics*, *34*(4). 320–326. https://doi.org/10.1123/jab.2017-0256

Vaiyyali, C. K. (2024, July). Technology in sports event management. *International Journal of Research Publication and Reviews*, *5*(7), 5035–5038. https://ijrpr.com/uploads/V5ISSUE7/IJRPR31900.pdf

Vannatta, S. (2011, August). Phenomenology and the question of instant replay: A crisis of the sciences? *Sport, Ethics and Philosophy 5*(3), 331–342. https://doi.org/10.1080/17511321.2011.602587

Vidal-Tomas, D. (2023). Blockchain, sport and fan tokens. *Journal of Economic Studies, 51*(1), 24–38. https://doi.org/10.1108/JES-02-2023-0094

Vina, A. (2025, March 19). The application and impact of AI in basketball and the NBA. *Ultralytics*. https://www.ultralytics.com/blog/application-and-impact-of-ai-in-basketball-and-nba

Walch, K. (2024, August 16). How AI is revolutionizing professional sports. *Forbes*. https://www.forbes.com/sites/kathleenwalch/2024/08/16/how-ai-is-revolutionizing-professional-sports/

Winand, M. & Fergusson, C. (2016, December 21). More decision-aid technology in sport? An analysis of football supporters' perceptions on goal-line technology. *Soccer and Society, 19*(7), 966–985. https://doi.org/10.1080/14660970.2016.1267629

Zhang, D., Liu, T., Li, S., Luo, H., Ding, G., Su, Z., Xie, C., Zhang, S., & Feng, P. (2022, November 1). Three-dimensional laser scanning for large-scale as-built surveying of the 2022 Beijing Winter Olympic Speed Skating Stadium: A case study. *Journal of Building Engineering, 59*. https://doi.org/10.1016/j.jobe.2022.105075

13

TECHNOLOGY AND SUSTAINABILITY IN SPORT AND RECREATION

Jacob Bustad and Gashaw Abeza

Introduction

Historically, sports and recreational activities have often relied on infrastructures, equipment, and travel, which often result in substantial environmental impacts. For example, facility construction and the production of sport and recreation equipment and apparel have contributed to waste, pollution, and carbon emissions (McCullough & Kellison, 2017), while outdoor recreation has affected ecosystems through overuse of trails, water pollution, and habitat disruption (Monz et al., 2013). However, growing awareness of climate change, biodiversity loss, and resource scarcity has driven innovation that reflects increased efforts to align sport, recreation, and leisure with environmental sustainability (Mascarenhas et al., 2021). Following Winter et al. (2019), the complexity of the relationship between sport, recreation, and the environment means that there is a need for different approaches to the delivery and distribution of recreation programs and services, as well as a shift in approach regarding the management and stewardship of recreation facilities and natural spaces.

This chapter focuses on the integration and uses of technology and technological innovations within sport and recreation management to support sustainable practices and policies in the community recreation and outdoor recreation sectors. This includes a discussion of the relationship between recreation, health, and the environment, an overview of specific technologies that have been incorporated into community and outdoor recreation activities, and an explanation of several broad frameworks that address environmental issues while also providing quality recreation experiences.

DOI: 10.4324/9781003588405-17

Background

Recreation and Public Health

Community and outdoor recreation serve as key aspects of public health, as recreation experiences provide formal and informal opportunities for physical activity and social interaction. Public parks, recreation centers, sports leagues, walking, hiking, and biking trails, and nature-based programming serve to facilitate the physical, emotional, and social well-being of individuals and communities (Frumkin et al., 2017; Wolsko et al., 2019). Community and outdoor recreation also play a broader role in promoting physical activity in an increasingly sedentary world, which in turn can address and prevent chronic diseases such as obesity, diabetes, and cardiovascular conditions (Omura et al., 2021). Accessible recreational facilities and services, such as public gyms, swimming pools, bike paths, and sports fields, can decrease barriers to physical activity participation, and the availability and proximity of recreational facilities often correlate with higher levels of physical activity among residents (Kaczynski & Henderson, 2007; Lee & Maheswaran, 2011).

Community and outdoor recreation are also important determinants of mental health, as participating in physical activities within outdoor or nature-based settings can reduce symptoms of stress, anxiety, and depression (Lackey et al., 2021). Sport and recreation in green spaces, including parks and trails, can therefore provide multiple mental health benefits, including opportunities for individuals to mentally disconnect from daily stressors and engage in positive self- and social reflection (Bratman et al., 2019).

Recreation and the Environment

Yet, while community and outdoor recreation provide physical, mental, and social benefits to individuals and communities, these activities can also have a range of negative environmental impacts. The ecological footprint of community-based recreation is therefore becoming an important consideration for planners, policymakers, and sport and recreation managers. As one of the most visible environmental impacts, recreational facilities often entail forms of land use change and habitat disruption, as parks, trails, fields, and other facilities necessitate the clearing and alteration of landscapes (Duerksen & Snyder, 2013). Within urban settings, the development of recreational facilities and artificial turf fields can serve to reduce biodiversity and existing ecological systems of plants and animals (Cheng, Hu & Reinhard, 2014). Facilities such as swimming pools, gyms, and community centers often require significant amounts of water, energy, and maintenance materials, including hazardous chemicals. The use of fertilizers and pesticides to maintain athletic fields can lead to chemical runoff, potentially harming nearby water bodies and aquatic ecosystems (Cheng et al., 2014).

In regard to natural and green spaces, Marzano and Dandy (2012) explain that frequent use of natural areas for informal recreation, including hiking and biking, can lead to trail erosion, soil compaction, and damage to native vegetation, cumulatively degrading ecosystems, especially in areas with high visitation. Litter and waste, such as water bottles and food packaging, is common in parks and at recreational events, and is exacerbated by increased visitation – for example, during the COVID-19 pandemic, increased visitors at national and state parks in the United States resulted in higher levels of waste, misuse of facilities and trails, and vandalism at those sites (Bustad et al., 2023). Recreational activities therefore contribute to pollution and resource overuse, including litter accumulation, noise pollution, and water pollution, and in turn these forms of environmental degradation can negatively influence the experience of parks and natural areas (Lyu et al., 2020). Further, emissions from transportation to and from recreational sites in car-dependent communities and regions represent a significant portion of recreation's carbon footprint (Gössling et al., 2019).

Sustainability and Technology

The relationship between recreation and the environment includes experiences and activities that have positive physical and mental health benefits for participants, and at the same time, these experiences and activities result in negative impacts on the environments in which they take place. However, while previous approaches to sport and recreation management have often viewed the natural environment as an unchanging element (Orr & Inoue, 2019), more recently these fields have exhibited an increased awareness and interest toward developing forms of sustainability that will support recreation and the environment simultaneously.

McCullough et al. (2020) explain that while sport has often been associated with "waves" of social change in regard to issues such as racial and gender equality, more recent changes to sport and recreation have also been characterized by an increased focus on the environmental impacts of activities and events. These authors contend that recent developments in the study and practice of sport and recreation have demonstrated several specific "green waves", including an initial awareness of the relationship between sport and the environment, a secondary shift from awareness to knowledge and education through additional research, and finally a transition from knowledge to actionable strategies that could be incorporated by sport and recreation organizations (McCullough et al., 2020, p. 10). As Blahna et al. (2020) state, the goal of "sustaining both recreation experiences and environmental conditions" has therefore become paramount for many sport and recreation managers in regard to developing, sharing, and implementing sustainable recreation practices, policies, and planning (p. 182). This chapter provides an overview of several key technologies and technological innovations that have been incorporated into efforts toward sustainable sport and recreation, as well as a discussion of how technology will be a key aspect of efforts toward sustainability within community and outdoor recreation moving forward.

Topical Exploration and Key Concepts

Sport and recreation organizations and managers seeking to integrate concerns for the environment have demonstrated several specific and related approaches to sustainability. This includes *environmental connectedness* (Beery & Wolf-Watz, 2014) in reference to increasing awareness of negative environmental impacts and also developing actions that can address and limit these impacts, and *environmental consciousness* (McCullough et al., 2020) in terms of a systematic and organizational culture of considering environmental issues as a priority along with the delivery and distribution of sport and recreation opportunities and experiences. These approaches have often been supported by a corresponding focus on recognizing that sport and recreation play an active role in shaping broader *ecosystem services* (Drius et al., 2019) comprised of interactions between communities and nature, as well as the *triple bottom line* (Carneiro et al., 2018) as an organizational goal for success according to fiscal, social, and environmental metrics. This section examines four specific examples of the utilization of technology within these broader approaches, in order to demonstrate particular areas where concerns for sustainability are evident within community and outdoor recreation: green infrastructure and facility design, resource consumption and conservation, data monitoring and environmental impacts, and sustainable apparel and equipment.

Green Infrastructure and Facility Design

Community and outdoor recreation infrastructure refers to elements within natural spaces such as parks, forests, and beaches, as well as elements of the built environment, including recreation centers, aquatic centers, sports complexes, and trails. These elements often involve the use of significant amounts of water, energy, and materials as part of their initial development and construction and subsequent maintenance and renovation. In turn, the integration of green infrastructure and sustainable facility design has been incorporated into community and outdoor recreation planning and policy in order to minimize environmental impacts while meeting recreational needs. Benedict and McMahon (2012) define green infrastructure as a strategically planned network of natural and semi-natural areas, designed to address a range of environmental factors and impacts, including water and air quality improvement, and support for preserving forms of biodiversity. Within recreational facility planning and design, green infrastructure is evident through modifications such as green roofs, permeable pavements, rain gardens (Zhou et al., 2018), and naturalized stormwater systems integrated into parks, trails, and recreation facilities (Martín Muñoz et al., 2024). Recreation planning has also incorporated green corridors and vegetated buffers that improve access to green and outdoor space for physical activity, while also reducing the need for artificial cooling within heat-prone areas (Pereira et al., 2024).

Resource Consumption and Conservation

The consumption of energy within recreation facilities via lighting, heating, and cooling and pool heating for aquatic centers has led to efforts toward more energy-efficient systems, such as lighting systems that use occupancy sensors to automatically dim or shut off lights. Other facilities have sought to harness energy through forms of technological innovation: in 2018, Eco Gym in Rochester, New York introduced 21 cardio machines that contributed energy from gym users toward facilities of energy consumption (Love, 2018), and in 2023, the city of Chilliwack, British Columbia implemented a system in which heat generated from the ice plant for the city's hockey rink is used to heat pools at a nearby aquatic facility (BC Hydro, 2023).

Water constitutes another resource consumed through recreation infrastructure, including irrigation for fields and parks and recreation centers and aquatic centers. Green infrastructure design can address these patterns of consumption through rainwater harvesting and greywater (wastewater without contamination) reuse systems, as these sources can be used for toilet facilities and landscape irrigation technologies such as drip systems and smart sprinkler systems (Livesley et al., 2021).

Data Monitoring and Environmental Impacts

As noted above, community and outdoor recreational activities offer significant physical, psychological, and social benefits, and at the same time, often result in negative impacts on the environment. Data monitoring technologies have served as vital tools in assessing and managing these impacts, and data monitoring systems can provide information that supports sustainability practices and policies within recreation organizations and providers (Manning et al., 2011). Environmental data monitoring systems, including remote air and water quality sensors, camera traps, GPS trackers, automated counters, and mobile app-based user reporting, can be integrated into recreation management strategies in order to collect, analyze, and interpret information about recreational activity and environmental consequences.

These systems can enable visitor tracking, allowing for real-time information on the number and distribution of users that can be analyzed to identify patterns of crowding and overuse, issues that can be addressed through strategies such as limiting access and trail rotation. Air and water quality sensors can help track levels of pollution from recreational vehicles and campsites, while camera traps and GPS trackers can monitor wildlife activity as a noninvasive method to understand human–wildlife interactions (Dertien et al., 2021). Data monitoring can also inform policies and planning for community and outdoor spaces and services, as agencies and organizations can use data to regulate capacity and identify priorities for further investment. However, data monitoring can also be limited through the risk of underutilization, as Biber (2013) explains that "even if effective monitoring data is gathered, it may not be used in decision making" (p. 68). The uses of data monitoring technology, therefore, also require institutional support in order to implement sustainable strategies.

Sustainable Apparel and Equipment

While community and outdoor recreation activities incorporate and celebrate elements of nature, the apparel and equipment used within these activities have their own environmental impacts. In particular, recreational gear such as apparel and footwear, tents and camping accessories, and other specialized equipment are most often designed for performance and durability, including the use of synthetic fabrics and composite materials that are resource-intensive and contribute to pollution and waste (Aall et al., 2023). For example, the manufacturing process for waterproof clothing designed for outdoor recreation has often involved chemical treatments with PFAS (per- and polyfluoroalkyl substances) and PFCs (perfluorinated chemicals) and toxic chemicals that can accumulate in the human body (Okamoto, 2025). The heavy use of plastics within recreational equipment has also been linked with waste and pollution in natural areas, including parks and waterways, with negative effects on ecologies and wildlife (Liu et al., 2022).

Lifecycle Assessment

In order to minimize and mitigate these environmental impacts, apparel and equipment manufacturers have utilized forms of technology that allow for greater evaluation of the type and amounts of resources used through the production process, in the form of lifecycle assessment (LCA). LCA measures can provide information that enable manufacturers to make changes within the production, distribution, and use cycle of apparel and equipment items in order to avoid a "take-make-waste" model (Lazarevic & Valve, 2017). Notable outdoor recreation brands including Patagonia and The North Face have incorporated

LCA measures in order to improve product sustainability: Patagonia's "Worn Wear" program seeks to extend the lifecycle of their products through in-store and online repair services (Michel et al., 2019), while The North Face's "Backyard Project" included a hooded sweatshirt that was produced as much as possible within a 150 mile radius of the company's Alameda California headquarters, supporting local farmers and garment workers while also limiting the carbon emissions of the supply chain (Rissanen & McQuillan, 2023). Other outdoor brands such as Helly Hansen have developed monomaterial products made from a single type of material without mixed fibers, which enables "closed-loop recycling" wherein the recycled materials are used to produce similar products (Helly Hansen, 2021). These efforts by manufacturers are also further supported by organizations that prioritize sustainability in the outdoor recreation industry, including the European Outdoor Group (EOG) and Outdoor Industry Association (OIA), by sharing LCA data and practices and policies that can reduce environmental impacts.

Practice in Action

Sustainability and Ski Tourism

As a specific sector of community and outdoor recreation, skiing and ski tourism constitute a "climate-sensitive tourism activity" (Bai & Zhang 2021, p. 2). Skiing can contribute to deforestation and wildlife disruption in alpine areas, while also requiring high amounts of water consumption for artificial snow production and energy consumption for ski lifts and guest services; at the same time, changing climates with warmer temperatures and reduced snowfall mean that ski resorts are vulnerable to shorter seasons and variable snowpack.

In response to these challenges, various organizations have developed initiatives that seek to promote sustainable winter sports through the protection of local ecosystems. These organizations often focus on particular aspects of the ski industry, including: reducing carbon emissions through renewable energy and the use of electric vehicles (EVs); utilizing improved and more efficient snowmaking systems, often involving reclaimed water; addressing waste reduction through zero-waste policies and recycling initiatives; preserving natural habitats and protecting biodiversity by minimizing the impact of ski trails; and engaging with skiers, local communities, and policymakers in advancing issues and ideas focused on sustainability.

Protect Our Winters (POW), an American-based nonprofit originally founded by professional snowboarder Jeremy Jones, has ties to winter sport athletes and ski communities. Their efforts include community campaigns and corporate partnerships that support policies to reduce carbon emissions and preserve winter sport environments. As part of the National Ski Areas Association in the United States, the Sustainable Slopes program enlists ski resorts and areas to incorporate sustainability into their operations through initiatives including environmental audits of resource consumption and waste, eco-friendly forms of design and construction, and climate education and advocacy. Finally, within the OIA, the SnowSport Forum emphasizes sustainable practices among ski resort operators regarding emissions reduction, as well as ski equipment manufacturers in implementing lifecycle assessments and recycling and reuse programs.

Trends and Current Issues

Along with specific technologies utilized within efforts to make community and outdoor recreation more sustainable, there are also several theories that have informed recent developments in sport and recreation management. Sport ecology, recreation ecology, and sustainable recreation and tourism are intersecting fields that seek to examine the relationship between human physical activity and natural environments, offering theoretical frameworks that better understand the dynamics of how sport and recreation impact the environment and also how environmental factors play a role in shaping practices, policies, and experiences of sport and recreation.

Sport Ecology

According to McCullough et al. (2020), the sport ecology framework asserts that physical activities, including sport and recreation, have a "bidirectional" relationship with the environment, and that these industries are increasingly characterized by efforts to address issues related to environmentalism and sustainability (p. 509). More specifically, sport ecology has often focused on aspects of *adaptation*, focused on how sport and recreation organizations can effectively modify their existing systems and sites in order to promote behavioral changes within athletes, fans, and industry professionals through policy design and innovative facility and event management (Orr & Schneider, 2018).

An example of the sport ecology approach is demonstrated through studies focused on carbon-neutral and climate-positive sport and recreation events, which seek to both limit negative impacts on the environment at both the local and global scales (Trendafilova & Ziakas, 2025). Moreover, erratic weather patterns involving extreme heat, wildfires, and reduced snowfall are forcing adaptations in several sports – as discussed above, ski resorts in Europe and North America have responded to shorter snow seasons by investing in energy- and resource-efficient infrastructure, including for artificial snow production. Endurance sports such as marathons are being rescheduled to avoid extreme heat periods and seasons, while event organizers are incorporating technology that allows for heat stress monitoring and hydration and safety protocols (Nowak et al., 2022).

Recreation Ecology

In comparison to sport ecology, the framework offered by recreation ecology has more often incorporated environmental and conservation science data and methods, in order to emphasize the impacts of recreational activities within natural and protected areas, including local, state, and national parks, forests, beaches, rivers, and wilderness zones (Monz et al., 2013). Stankey (2020) explains that two important aspects within recreation ecology include the recreation opportunity spectrum (ROS) model, focused on how different environments and landscapes can be categorized to guide specific types and capacities of recreational activities based on ecological sensitivity, and the limits of acceptable change (LAC) model, which utilizes environmental data to establish particular thresholds of ecological damage that can be used to adaptively manage human activity within natural and wilderness areas. Each of these models requires the use of technology and technological innovation, in regard to data monitoring, tracking, and analysis of data to effectively identify and manage areas at risk of environmental degradation.

Recreation ecology has also contributed to the development of other strategies that align with sustainability, such as the Leave No Trace initiative for hikers, bikers, swimmers, surfers, campers, and other participants in outdoor recreation activities (Cole, 2021). Leave No Trace incorporates seven specific principles that seek to increase awareness of the effects of human–nature interactions, in order to reduce the footprint of recreational activities by preventing or limiting negative impacts, including soil erosion, damage to vegetation, wildlife disturbance, and decreases to water quality (Backman et al., 2018). The increase in visitors at many protected areas, such as national and state parks, forests, and beaches, during the COVID-19 pandemic underscored the importance of Leave No Trace programs, as effective implementation of these programs can educate users and provide a basis for behavior change toward minimizing and mitigating environmental impacts (Ferguson et al., 2022). Along with efforts to reduce the negative effects of recreational activities, the recreation ecology approach can also contribute toward the integration of recreation with environmental advocacy and activism, as with organized beach and shoreline clean events (Wyles et al., 2017).

Sustainable Tourism

Tourism serves as a key aspect of many recreational activities, as individuals, families, and groups often travel locally, regionally, and globally in order to facilitate particular recreation experiences. At the same time, tourism constitutes a major challenge for sustainability, as it usually involves the consumption of fossil fuels for auto and air travel and the potential for increased waste and pollution, meaning that the growth of the tourism industry has, in many cases, become a problem for local environments (Higgins-Desbiolles, 2018). As with sport ecology and recreation ecology, sustainable tourism has developed as a specific field of study and practice relevant to sport and recreation managers, as it aims to balance the environmental, social, and economic dimensions of human activity (Harris et al., 2012).

The engagement with sustainability as an integral part of tourism planning and policies has contributed to the development of alternative types of tourism, including ecotourism and adventure tourism. Ecotourism is often practiced in areas of ecological significance, including mountains, rainforests, and coral reefs, and offers experiences that preserve the ecological integrity of these destinations – for example, ecotourism is demonstrated through activities such as wildlife viewing, birdwatching, and guided hiking, biking, and swimming tours. Most importantly, ecotourism activities prioritize long-term sustainability by minimizing environmental impacts while also engaging with and promoting local community development, in contrast to traditional forms of tourism (Khaledi Koure et al., 2023). Adventure tourism, involving specialized activities such as rafting, rock climbing, ziplining, and sky diving, often seeks to structure human–nature interactions in a manner that protects environments while also providing unique experiences (Hanna et al., 2019). Ecotourism and adventure tourism are increasingly characterized by technological innovations that can improve the experience for tourists, providers, and local governments alike by providing connectivity and quick access to information for tourists while also supporting tourism operations through monitoring of visitor and environmental data (O'Beirne, 2025).

Summary

As this chapter has discussed, physical activity and recreation can provide physical and social health benefits, and community and outdoor recreation promote healthy lifestyles through experiences in and with nature. However, human activity within natural spaces and landscapes often results in adverse environmental impacts, and there has been a growing concern within sport and recreation management to develop theories and strategies that can reduce these impacts while enhancing the long-term viability of recreational activities and environmental settings.

Following a discussion of the relationship between recreation, health, and the environment, this chapter examines specific technologies and forms of technological innovation that are shaping recreational sites and services, as well as theoretical and managerial approaches that provide guidance toward identifying and reaching sustainability goals. The continued incorporation of technology will therefore be a key aspect of sustainability initiatives for community and outdoor recreation organizers moving forward, in order to ensure the provision of quality recreation experiences that support local ecological conditions.

Key Resources

- **Green Sports Alliance**: https://greensportsalliance.org/
- **Outdoor Industry Association**: https://outdoorindustry.org/
- **Sports for Climate Action Framework**: https://unfccc.int/climate-action/sectoral-engagement/sports-for-climate-action
- **Leave No Trace Center for Outdoor Ethics**: https://lnt.org/

References

Aall, C., Klepp, I. G., Engeset, A. B., Skuland, S. E., & Støa, E. (2023). Leisure and sustainable development in Norway: Part of the solution and the problem. In L. Mansfield & B. Wheaton (Eds.), *Leisure and the Politics of the Environment* (pp. 71–94). Routledge.

Backman, C. L., Vaske, J. J., Lawhon, B., Vagias, W. M., Newman, P., Coulson, E., & Taff, B. D. (2018). Visitors' views of leave no trace principles across a national park, a national forest, and three state parks. *Journal of Park and Recreation Administration, 36*(4).

Bai, Z., & Zhang, Y. (2021). Sustainability of ski tourism in China: An integrated model of skiing tourists' willingness to pay for environmental protection. *Sustainability, 13*(16), 8883–8899.

BC Hydro. (2023, September 29). Chilliwack ice arena heats pools, cuts greenhouse gases. *BC Hydro*. https://www.bchydro.com/news/conservation/2023/chilliwack-rink-heats-pools.html

Beery, T. H., & Wolf-Watz, D. (2014). Nature to place: Rethinking the environmental connectedness perspective. *Journal of Environmental Psychology, 40*, 198–205.

Benedict, M. A., & McMahon, E. T. (2012). *Green infrastructure: linking landscapes and communities*. Island Press.

Biber, E. (2013). The challenge of collecting and using environmental monitoring data. *Ecology and Society, 18*(4), 68–81.

Blahna, D. J., Kline, J. D., Williams, D. R., Rogers, K., Miller, A. B., McCool, S. F., & Valenzuela, F. (2020). Integrating social, ecological, and economic factors in sustainable recreation planning and decision-making. In S. Selin, L. K. Cerveny, D. J. Blahna, & A. B. Miller (Eds.), *2020. Igniting Research for Outdoor Recreation: Linking Science, Policy, and Action* (257, pp. 987, 173–188). US Department of Agriculture, Forest Service, Pacific Northwest Research Station.

Bratman, G. N., Anderson, C. B., Berman, M. G., Cochran, B., De Vries, S., Flanders, J., Daily, G. C. (2019). Nature and mental health: An ecosystem service perspective. *Science Advances, 5*(7), 5–19.

Bustad, J. J., Clevenger, S. M., & Rick, O. J. (2023). COVID-19 and outdoor recreation in the post-anthropause. *Leisure Studies, 42*(1), 85–99.

Carneiro, M. J., Breda, Z., & Cordeiro, C. (2018). Sports tourism development and destination sustainability: the case of the coastal area of the Aveiro region, Portugal. In B. Moyle, T. Hinch, & J. Higham (Eds.), *Sport Tourism and Sustainable Destinations* (pp. 143–172). Routledge.

Cheng, H., Hu, Y., & Reinhard, M. (2014). Environmental and health impacts of artificial turf: A review. *Environmental Science & Technology, 48*(4), 2114–2129.

Cheng, Z., Lee, J. H., Day, S. D., & Strahm, B. D. (2014). Soil and water quality in urban athletic fields: Impacts of turf management practices. *Urban Forestry & Urban Greening, 13*(4), 714–724.

Cole, D. N. (2021, January). Reflections on the early history of recreation ecology. *Parks Stewardship Forum, 37*(2), 379–400.

Dertien, J. S., Larson, C. L., & Reed, S. E. (2021). Recreation effects on wildlife: A review of potential quantitative thresholds. *Nature Conservation*, *44*, 51–68.
Drius, M., Bongiorni, L., Depellegrin, D., Menegon, S., Pugnetti, A., & Stifter, S. (2019). Tackling challenges for Mediterranean sustainable coastal tourism: An ecosystem service perspective. *Science of the Total Environment*, *652*, 1302–1317.
Duerksen, C., & Snyder, C. (2013). *Nature-Friendly Communities: Habitat Protection and Land Use Planning*. Island Press.
Ferguson, M. D., McIntosh, K., English, D. B., Ferguson, L. A., Barcelona, R., Giles, G., … & Leberman, M. (2022). The outdoor renaissance: Assessing the impact of the COVID-19 pandemic upon outdoor recreation visitation, behaviors, and decision-making in New England's national forests. *Society & Natural Resources*, *35*(10), 1063–1082.
Frumkin, H., Bratman, G. N., Breslow, S. J., Cochran, B., Kahn Jr, P. H., Lawler, J. J., … Wood, S. A. (2017). Nature contact and human health: A research agenda. *Environmental Health Perspectives*, *125*(7), 075001.
Gössling, S., Scott, D., & Hall, C. M. (2019). The decarbonisation of travel: Key issues. *Journal of Sustainable Tourism*, *27*(5), 665–678.
Hanna, P., Wijesinghe, S., Paliatsos, I., Walker, C., Adams, M., & Kimbu, A. (2019). Active engagement with nature: Outdoor adventure tourism, sustainability and wellbeing. *Journal of Sustainable Tourism*, *27*(9), 1355–1373.
Harris, R., Williams, P., & Griffin, T. (Eds.). (2012). *Sustainable Tourism*. Routledge.
Helly Hansen. (2021, March 10). Mono material: Singular & circular design. *Helly Hansen*. https://www.hellyhansen.com/journal/mono-material-singular-and-circular-design
Higgins-Desbiolles, F. (2018). Sustainable tourism: Sustaining tourism or something more? *Tourism Management Perspectives*, *25*, 157–160.
Kaczynski, A. T., & Henderson, K. A. (2007). Environmental correlates of physical activity: A review of evidence about parks and recreation. *Leisure Sciences*, *29*(4), 315–354.
Khaledi Koure, F., Hajjarian, M., Hossein Zadeh, O., Alijanpour, A., & Mosadeghi, R. (2023). Ecotourism development strategies and the importance of local community engagement. *Environment, Development and Sustainability*, *25*(7), 6849–6877.
Lackey, N. Q., Tysor, D. A., McNay, G. D., Joyner, L., Baker, K. H., & Hodge, C. (2021). Mental health benefits of nature-based recreation: A systematic review. *Annals of Leisure Research*, *24*(3), 379–393.
Lazarevic, D., & Valve, H. (2017). Narrating expectations for the circular economy: Towards a common and contested European transition. *Energy Research & Social Science*, *31*, 60–69.
Lee, A. C., & Maheswaran, R. (2011). The health benefits of urban green spaces: A review of the evidence. *Journal of Public Health*, *33*(2), 212–222.
Liu, H., Pan, H., Chu, P., & Huo, D. (2022). Impact of plastic pollution on outdoor recreation in the existence of bearing capacity and perspective management. *Environmental Research*, *214*, 113819.
Livesley, S. J., Marchionni, V., Cheung, P. K., Daly, E., & Pataki, D. E. (2021). Water smart cities increase irrigation to provide cool refuge in a climate crisis. *Earth's Future*, *9*(1), 1–6.
Love, T. (2018, May 23). This new eco-gym is powered by your workout. *Green Matters*. https://www.greenmatters.com/community/2018/03/01/2b7lvo/eco-gym-workouts
Lyu, S., Zhang, C., & Zhang, H. (2020). Assessing visitors' perceptions of environmental impacts of tourism and recreation in urban parks: A case study in Beijing. *Urban Forestry & Urban Greening*, *52*, 126688.
Manning, R., Valliere, W., Anderson, L., McCown, R. S., Pettengill, P., Reigner, N., & van Riper, C. (2011). Defining, measuring, monitoring, and managing the sustainability of parks for outdoor recreation. *Journal of Park and Recreation Administration*, *29*(3), 24–37.

Martín Muñoz, S., Elliott, S., Schoelynck, J., & Staes, J. (2024). Urban storm water management using nature-based solutions: A review and conceptual model of floodable parks. *Land*, *13*(11), 1858.

Marzano, M., & Dandy, N. (2012). Recreationist behaviour in forests and the disturbance of wildlife. *Biodiversity and Conservation*, *21*(11), 2967–2986.

Mascarenhas, M., Pereira, E., Rosado, A., & Martins, R. (2021) How has science highlighted sports tourism in recent investigation on sports' environmental sustainability? A systematic review. *Journal of Sport & Tourism*, *25*(1), 42–65. https://doi.org/10.1080/14775085.2021.1883461

McCullough, B. P., & Kellison, T. B. (2017). An introduction to environmental sustainability and sport. In B.P. McCullough & T.B. Kellison (Eds.), *Routledge Handbook of Sport and the Environment* (pp. 3–10). Routledge.

McCullough, B. P., Orr, M., & Kellison, T. (2020). Sport ecology: Conceptualizing an emerging subdiscipline within sport management. *Journal of Sport Management*, *34*(6), 509–520.

Michel, G. M., Feori, M., Damhorst, M. L., Lee, Y. A., & Niehm, L. S. (2019). Stories we wear: Promoting sustainability practices with the case of Patagonia. *Family and Consumer Sciences Research Journal*, *48*(2), 165–180.

Monz, C. A., Pickering, C. M., & Hadwen, W. L. (2013). Recent advances in recreation ecology and the implications of different relationships between recreation use and ecological impacts. *Frontiers in Ecology and the Environment*, *11*(8), 441–446.

Nowak, A. S., Kennelley, G. E., Krabak, B. J., Roberts, W. O., Tenforde, K. M., & Tenforde, A. S. (2022). Endurance athletes and climate change. *The Journal of Climate Change and Health*, *6*, 100118.

O'Beirne, C. (2025). Technology and adventure tourism management. In G. Pomfret, A. Doran, & C. Cater (Eds.), *Routledge International Handbook of Adventure Tourism* (pp. 331–346). Routledge.

Okamoto, K. (2025, January 13). PFAS bans are coming for clothing. Here's what you need to know. *New York Times*. https://www.nytimes.com/wirecutter/reviews/pfas-bans-for-clothing/

Omura, J. D., Whitfield, G. P., Chen, T. J., Hyde, E. T., Ussery, E. N., Watson, K. B., & Carlson, S. A. (2021). Surveillance of physical activity and sedentary behavior among youth and adults in the United States: History and opportunities. *Journal of Physical Activity and Health*, *18*(S1), S6–S24.

Orr, M., & Inoue, Y. (2019). Sport versus climate: Introducing the climate vulnerability of sport organizations framework. *Sport Management Review*, *22*(4), 452–463.

Orr, M., & Schneider, I. (2018). Substitution interests among active-sport tourists: The case of a cross-country ski event. *Journal of Sport & Tourism*, *22*(4), 315–332.

Pereira, C., Flores-Colen, I., & Mendes, M. P. (2024). Guidelines to reduce the effects of urban heat in a changing climate: Green infrastructures and design measures. *Sustainable Development*, *32*(1), 57–83.

Rissanen, T., & McQuillan, H. (2023). *Zero Waste Fashion Design*. Bloomsbury Publishing.

Stankey, G. H. (2020). The recreation opportunity spectrum and the limits of acceptable change planning systems: A review of experiences and lessons. *Ecosystem Management*, 173–188.

Trendafilova, S., & Ziakas, V. (2025). Advancing sustainable sport event management and sport ecology: The missing links. *Journal of Policy Research in Tourism, Leisure and Events*, 1–16. https://www.tandfonline.com/doi/full/10.1080/19407963.2025.2459732

Winter, P. L., Crano, W. D., Basáñez, T., & Lamb, C. S. (2019). Equity in access to outdoor recreation—Informing a sustainable future. *Sustainability*, *12*(1), 124–150.

Wolsko, C., Lindberg, K., & Reese, R. (2019). Nature-based physical recreation leads to psychological well-being: Evidence from five studies. *Ecopsychology*, *11*(4), 222–235.

Wyles, K. J., Pahl, S., Holland, M., & Thompson, R. C. (2017). Can beach cleans do more than clean-up litter? Comparing beach cleans to other coastal activities. *Environment and Behavior*, *49*(5), 509–535.

Zhou, L., Shen, G., Woodfin, T., Chen, T., & Song, K. (2018). Ecological and economic impacts of green roofs and permeable pavements at the city level: The case of Corvallis, Oregon. *Journal of Environmental Planning and Management*, *61*(3), 430–450.

PART V
Conclusion

14

A REFLECTION ON TECHNOLOGY, SPORT, AND RECREATION FRAMEWORK

Gashaw Abeza and Jacob Bustad

A Reflection on Technology, Sport, and Recreation Framework

The Technology–Sport–Recreation Framework (TSRF), introduced in Chapter 1, offers a conceptual approach to understanding the intersection of technology, sport, and recreational activities. In recent years, the convergence of the technology, sport, and recreation domains has rapidly evolved, driven by advancements such as artificial intelligence (AI), wearable technology, virtual and augmented reality (VR/AR), big data analytics, and digital platforms. These technological innovations reshaped participation, performance enhancement, immersive experiences, and community engagement in both sport and recreational contexts. As society becomes increasingly digitized and interconnected, an understanding of how technology interacts with and transforms sport and recreation is crucial for leveraging technological innovations to enhance individual health and performance, facilitate community connections, foster inclusivity, and achieve broader sustainability goals. Anchored by practical insights presented across the chapters in this book, this concluding chapter reflects on TSRF by critically evaluating its strengths, limitations, and potential future directions, highlighting how the framework serves as a foundational model to guide scholarly inquiries and practical implementations within the multifaceted and rapidly evolving field of sport, recreation, and technology. The chapter synthesizes insights from diverse technological applications and empirical contexts, offering a reflection on the TSRF's current impact and future potential.

DOI: 10.4324/9781003588405-19

Strengths of the TSRF

A fundamental strength of the TSRF lies in its clear delineation of the Technology, Sport, and Recreation Ecosystem, comprehensively integrating five core components (see Figure 14.1): technological infrastructure, digital innovations, sport and recreation activities, stakeholders, and management and governance. Chapter 1 effectively establishes this ecosystem, emphasizing the interconnectedness and interdependence of these components and highlighting their collective role in shaping user experiences, outcomes, and broader societal impacts.

The practical insights across the preceding chapters underscore the strength of this ecosystem approach. Chapters examining wearable technologies (Chapter 2), gamification strategies (Chapter 3), and sport analytics (Chapter 4) substantiate the TSRF's applicability in performance enhancement contexts. These chapters highlight the transformative potential of technology in optimizing individual performance, motivation, and behavioral outcomes. Furthermore, chapters exploring immersive technologies such as AI (Chapter 5), AR and VR (Chapter 6), esports platforms (Chapter 7), digital transformation in traditional sport (Chapter 8), and innovative sport-tech startups (Chapter 9) illustrate technology's capacity to enhance immersive experiences and reshape traditional sport practices.

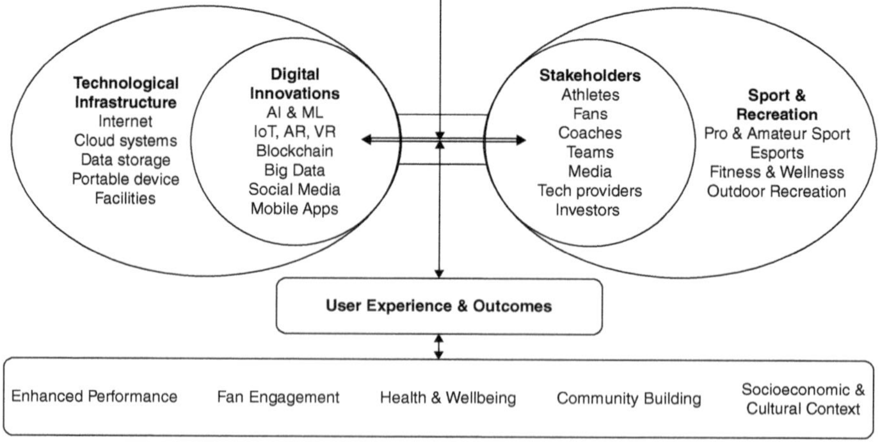

FIGURE 14.1 Technology, Sport, and Recreation Ecosystem (same as Figure 14.1 in Chapter 1).

Another significant strength is the framework's emphasis on digital communities, as discussed in Chapter 10. This layer highlights how digital platforms facilitate global connectivity, transforming sport fandom and community participation. Chapter 11 further enriches this layer, addressing the critical role of technology in adaptive sport, which promotes inclusivity and accessibility, particularly for individuals with disabilities. The exploration of various digital communities such as esports forums, fantasy sport networks, and virtual fitness groups underscores the TSRF's ability to capture the diverse forms of digital community engagement, addressing how digital technologies have redefined collective identity, loyalty, and social connectivity within sport and recreational contexts.

The TSRF's smart environments and sustainability layer, articulated through Chapters 12 and 13, demonstrates technology's role in optimizing venue management, event logistics, resource conservation, and environmental sustainability. These chapters collectively emphasize the position of technological innovations with global sustainability objectives, reflecting the framework's comprehensive consideration of societal and environmental implications. Specifically, the examination of technologies such as AI, blockchain, digital twins, and wearable devices within venue management and event logistics illustrates the practical applicability and relevance of the TSRF in addressing contemporary sustainability challenges and enhancing overall event experiences.

Additionally, the TSRF effectively synthesizes theoretical foundations with practical applications. By integrating scholarly research, practical case studies, and industry insights across the preceding 13 chapters, the framework demonstrates its use for both academia and practitioners. For instance, insights provided by sport-tech startups highlight how innovative business models and technological advancements not only reshape industry practices but also contribute significantly to public health and cultural perceptions of sport. Moreover, the critical examinations of ethical considerations such as data privacy, algorithmic bias, and equity issues across multiple chapters underline the framework's attention to responsible and ethical technology deployment.

Limitations of the TSRF

Despite its strengths, the TSRF has some limitations. First, the current practical insights predominantly emerge from contexts within technologically advanced regions, limiting the broader applicability of the framework globally. Chapters primarily emphasize examples and cases from developed nations, potentially overlooking diverse technological integration challenges and opportunities in less-developed or/and emerging markets. A broader representation of geographical contexts and socioeconomic conditions could provide deeper insights into how technological innovations operate differently across various settings. Second, although various chapters touch upon ethical dimensions like data privacy, algorithmic bias, and equity, there is room for more systematic treatment of socio-ethical considerations. For instance, issues regarding surveillance, fairness in digital engagement, and the socioeconomic divide in access to advanced technologies require deeper exploration to enhance responsible technological deployment across all TSRF layers. Furthermore, a much deeper examination of the implications of data governance, intellectual property rights, and cybersecurity within sport and recreation contexts would enhance the robustness of the framework.

Additionally, the TSRF currently lacks a detailed exploration of the potential negative consequences of technological adoption, such as digital dependency, diminished physical activity in favor of digital interactions, and the potential of increased screen time negatively impacting mental and physical health. Addressing these concerns within the framework would be areas that should be addressed by future research works, by acknowledging both the positive and potential adverse effects of technology integration. Moreover, while building the foundation is essential, given the rapid pace of technological evolution, maintaining the framework's relevance and accuracy necessitates frequent revisions and adaptive measures. The rapid advancements in blockchain, digital twins, extended reality, and other emerging technologies (e.g., Chapter 12) demand continuous refinement to accommodate new developments and unforeseen implications. The dynamic nature of technological advancements also poses challenges for policy and regulatory adaptation, underscoring the need for the TSRF to adapt with and incorporate mechanisms for timely policy updates and adaptations.

Future Directions for Research and Practice

With the rapid evolution of the convergence of the technology, sport, and recreation domains, future research and practice can build upon the existing foundations of the TSRF by exploring several promising avenues. First, extending the empirical evidence base through international comparative research would enhance understanding of the differential impacts and contextual variations of technological innovations across different global settings. Empirical studies comparing technology integration in sport and recreation across economically diverse regions could illuminate unique adoption challenges, successes, and best practices, thereby enriching the TSRFs global relevance. Second, there is a need for exploration of ethical and governance frameworks to manage the socio-ethical implications of technology use. An in-depth empirical study on ethical governance models, regulatory compliance, data privacy standards, and equitable access to technologies can inform policymaking and industry practices. Establishing empirically informed guidelines and best practices to ensure vethical and responsible deployment of technologies will bolster the TSRF's robustness and applied value.

Third, interdisciplinary collaboration will be a critical area for future development. Engaging researchers and practitioners from diverse fields, including technology, sport management, sociology, business, ethics, public health, and environmental science, could generate holistic perspectives on the convergence of the technology, sport, and recreation domains. Such collaborations would enable more nuanced and balanced analyses, narrowing as well as bridging the gap between technological innovations, societal expectations, ethical considerations, and practical implementations. Fourth, longitudinal research assessing the implications of sustained technological adoption in sport and recreation over an extended period is essential. Exploring the prolonged effects of wearables, gamification, and immersive technologies on individuals' physical health, mental well-being, and social behaviors would provide insights for sustainable technological integration. Such studies would inform effective and adaptive strategies to mitigate potential negative outcomes, such as technology dependency and reduced physical activity, ensuring beneficial long-term impacts.

Additionally, future research should focus on anticipating technological advancements and their associated impacts on the TSRF. Proactive identification of emerging technologies, such as advanced robotics, enhanced biometric sensors, and next-generation immersive platforms, could ensure that the TSRF remains adaptive and forward-looking. Continuous updates, periodic reviews, and iterative refinements of the framework should be essential, ensuring its sustained relevance in rapidly evolving technological landscapes. Furthermore, research on user acceptance, engagement, and behavioral adaptation to technological interventions in sport and recreation contexts warrants an in-depth investigation. Understanding psychological and sociological factors influencing user adoption, continued use, and potential resistance would facilitate more effective design and implementation of technological solutions. User-centric research approaches could help optimize user experiences, maximize beneficial outcomes, and minimize potential barriers to technology adoption. Finally, exploring innovative and sustainable business models driven by technological advancements represents a vital research direction. Investigations into economic impacts, market dynamics, and commercialization strategies related to sport-tech startups, digital platforms, and sustainability initiatives could identify, and thereby enhance, successful frameworks for sustainable economic growth and responsible technological commercialization within the sport and recreation sectors.

Summary

The Technology–Sport–Recreation Framework (TSRF), grounded in practical evidence across various chapters, offers a comprehensive lens to understand technology's transformative impacts on individual, community, and societal levels. Its systematic approach effectively categorizes technological interventions, clarifying their multidimensional effects and articulating the interconnectedness among the components of technological infrastructure, digital innovations, sport and recreation activities, stakeholders, and management and governance, which are elements of the TSR framework. Addressing the identified limitations through expanded empirical contexts, deeper ethical considerations, enhanced interdisciplinary collaboration, and continuous adaptation to rapid technological advancements will significantly strengthen the TSRF's applied value. Emphasizing global inclusivity, equitable access to technologies, and strong ethical governance models will ensure responsible and beneficial deployment across diverse settings.

It would be imperative for future research works to focus on longitudinal and comparative studies to provide deeper, contextualized insights into long-term implications, user engagement dynamics, and sustainable business practices. The proactive integration of emerging technologies such as advanced robotics and next-generation immersive platforms will further refine the TSRF's adaptability and forward-thinking nature. In due course, the TSRF provides a robust conceptual and practical foundation essential for leveraging technological innovations to optimize sport and recreation experiences. By nurturing ongoing interdisciplinary collaboration, the framework can effectively address evolving challenges, maximize positive outcomes, and minimize potential risks associated with technological integration. The continuous refinement and expansion of the TSRF will support sustained empirical advancements in the topic areas of individual health and performance, strengthen community engagement, and contribute positively to global sustainability and societal development goals.

INDEX

Pages in *italics* refer to figures and pages in **bold** refer to tables.

For Product Safety Concerns and Information please contact our EU
representative GPSR@taylorandfrancis.com
Taylor & Francis Verlag GmbH, Kaufingerstraße 24, 80331 München, Germany

www.ingramcontent.com/pod-product-compliance
Ingram Content Group UK Ltd.
Pitfield, Milton Keynes, MK11 3LW, UK
UKHW020146200926
475702UK00016B/2576
9781032961804